Alan Preuß

Development of high-temperature superconductor cables for high direct current applications

Karlsruher Schriftenreihe zur Supraleitung Band 031

HERAUSGEBER

Prof. Dr. Tabea Arndt
Prof. Dr. rer. nat. Bernhard Holzapfel
Prof. Dr. rer. nat. Sebastian Kempf
Prof. Dr.-Ing. Mathias Noe

Eine Übersicht aller bisher in dieser Schriftenreihe erschienenen Bände finden Sie am Ende des Buches.

Development of high-temperature superconductor cables for high direct current applications

by
Alan Preuß

KIT Scientific Publishing

Karlsruher Institut für Technologie
Institut für Technische Physik

Development of high-temperature superconductor cables for high direct current applications

Zur Erlangung des akademischen Grades eines Doktor-Ingenieurs von der KIT-Fakultät für Elektrotechnik und Informationstechnik des Karlsruher Instituts für Technologie (KIT) genehmigte Dissertation

von M. Sc. Alan Preuß

Tag der mündlichen Prüfung: 24. Oktober 2019
Hauptreferent: Prof. Dr.-Ing. Mathias Noe
Korreferent: Prof. Antonio Morandi, Ph.D.

Impressum

KIT Scientific Publishing

Karlsruher Institut für Technologie (KIT)
KIT Scientific Publishing
Straße am Forum 2
D-76131 Karlsruhe

www.ksp.kit.edu

Print on Demand 2021 – Gedruckt auf FSC-zertifiziertem Papier

ISSN 1869-1765
ISBN 978-3-7315-1041-3
DOI 10.5445/KSP/1000122861

1 Acknowledgments

First of all, I would like to thank all my colleagues at the Institute for Technical Physics (ITEP) at the Karlsruhe Institute of Technology (KIT) for their support during my time at the institute.

In particular I would like to thank Prof. Dr-Ing. Mathias Noe for the continuous support and constructive feedback I received right from the beginning of my Ph.D. I especially appreciate the dedication to teaching the Ph.D. students in the many seminars and workshops that were organized at the ITEP under the supervision of Prof. Noe.

My special gratitude goes to my supervisors in the Fusion magnet group Dr. Walter Fietz and Dr. Michael Wolf for enabling this dissertation and always having an open ear for problems and ideas. Thank you for encouraging and enabling the participation in seminars, workshops and conferences as well as proofreading my publications and this dissertation. The pleasant atmosphere within the fusion magnet group was one of the reason I started this thesis and is something I will remember fondly.

I would also like to thank Paul Wagner-Nagy and Thomas Bast with whom I spent many unforgettable hours on the CroCo manufacturing. Thank you for being optimists when I needed it most.

This dissertation would not have been possibly without the outstanding work of Christian Lange, Frank Gröner, Mathias Heiduk, Markus Hollik, Mirko Gehrlein, Ralph Lietzow, Ralf Müller, Uwe Mirasch, Andreas Kienzler and Simeon Eckerle.

My gratitude also have Dr. Alexandra Jung and Karoline Altmann for their help in preparing cross sections of the many failed samples and helping to understand what went wrong.

Finally, I want to thank my parents Ines and Ulrich for setting me on the path that eventually lead to the writing of this dissertation and their support throughout my studies. Thanks to my sisters Lena and Janne and to my fiance Jeanne for listening in tough times.

Alan Preuß

31.05.2019

2 Zusammenfassung

Der Schwerpunkt dieser Arbeit liegt auf der Entwicklung, Herstellung und Prüfung der supraleitenden Elemente eines 35 kA HTS DC Kabeldemonstrators. Drei wichtige HTS-Hochstromleiterkonzepte (Stapel, Roebel, Cable on Round Core (CORC)) wurden analysiert und das HTS Cross Conductor (HTS CroCo) Stapelkonzept ausgewählt. Für Hochstromanwendungen wurde ein Designprozess für HTS-Gleichstromkabel entwickelt. Basierend auf dem Designprozess wurde ein 100 kA Kabeldesign auf Basis von 36 CroCos entwickelt und anschließend auf einen 35 kA Kabeldemonstrator mit 12 CroCos herunterskaliert. Die Länge des Kabeldemonstrators wurde auf 3,6 m und die Betriebstemperatur auf 77 K festgelegt.

Vor der CroCo-Herstellung wurde eine Studie über das Degradationsverhalten der kritische Stromdichte von HTS-Bändern der zweiten Generation unter thermischer Belastung durchgeführt. Diese Untersuchungen sind notwendig, da HTS Bänder während der Herstellung oft thermischen Belastungen ausgesetzt sind, z.B. durch Löten. Ein Verständnis des genauen Degradationsverhaltens kann daher zur Optimierung von Fertigungsprozessen beitragen. Die Untersuchung ergab ein vorhersagbares und wiederholbares Degradationsverhalten, das durch die Sauerstoffverarmung des Kristalls verursacht zu werden scheint. Es wurde gezeigt, dass Zinn-Silber-Lote eine kostengünstige Alternative zu bleihaltigen Loten in der CroCo-Herstellung sein könnten. Die Ergebnisse der Untersuchung sind in [Pre+18] veröffentlicht.

Für die Herstellung der CroCos wurden Supraleiter zweier Hersteller auf ihre Kompatibilität mit dem CroCo-Herstellungsprozess untersucht. Es wurde festgestellt, dass Supraleiter eines Herstellers grundsätzlich kompatibel sind, aber starke Tendenzen zu einer ungleichen Verteilung von galvanisiertem Kupfer aufweisen. Supraleiter des anderen Herstellers erwiesen sich zu dem Zeitpunkt als nicht kompatibel, sind aber vielversprechende Kandidaten für zukünftige Arbeiten.

Der CroCo-Herstellungsprozess wurde weiterentwickelt, um eine runde äußere Lötmatrix zu integrieren. Darüber hinaus wurde der Herstellungsprozess angepasst, um Supraleiter mit minimalem galvanisiertem Kupfer (ca. 5-10 µm) zu erlauben, was dazu beiträgt, die Kosten für die Supraleiter zu senken. Um diese Veränderungen zu ermöglichen, wurde eine neue CroCo-Fertigungsmaschine eingerichtet. Die 12 CroCos, die im Demonstrationskabel verwendet wurden, wurden in 8 Produktionsläufen produziert. Aus jedem CroCo-Produktionslauf wurden ein oder zwei der gewünschten 3,6 m langen CroCos geschnitten. Der kritische Strom jedes der 12 CroCos wurde unter Eigenfeldbedingungen bei 77 K gemessen und variierte zwischen 2890 A und etwa 3680 A. Die relativ große Variation ist auf die Optimierung zahlreicher Fertigungsschritte und eine steigende Leistung der Supraleiter zurückzuführen. Es wurde gezeigt, dass der kritische Strom eines CroCo zuverlässig berechnet werden kann, basierend auf dem durchschnittlichen kritischen Strom der Supraleiter, die zur Herstellung des CroCo verwendet werden. Lötverbindungen wurden verwendet, um den CroCos mit normal leitenden Kupferkabeln zu verbinden. Der Widerstand dieser Verbindungen betrug im Durchschnitt 200 nΩ und mit einer Standardabweichung von 43 nΩ bei 77 K. Dies würde zu einer Wärmelast von ca. 0,7 mW pro Anschluss bei einem Betriebsstrom von 3500 A führen. Dies sollte für die meisten Anwendungen ausreichend niedrig sein.

Für den Test des Kabeldemonstrators wurde ein 6 m langer, 1 m breiter und 0,8 m hoher Kryostat entwickelt und gebaut. Das Demonstrationskabel selbst besteht aus den 12 CroCos mit je 22 6 mm breiten und 10 4 mm breiten Supraleiter Bändern. Die CroCos sind auf einem Aluminiumkern mit einem Durchmesser von 110 mm montiert und parallel geschaltet.

Während der Messreihe erreichte der Kabeldemonstrator einen stationären Strom von 34 kA und 35 kA für kurze Zeit. Eine Analyse der Stromverteilung während des Kabeltests ergab eine relativ große Streuung der Stromverteilung über die CroCos, die wahrscheinlich dadurch verursacht wurde, dass Wasser zwischen den Kontakten kondensierte und somit den Übergangswiderstand erhöhte. Ein CroCo wurde nach dem Test vom Demonstrator demontiert, um sicherzustellen, dass keine Degradation, z.B. durch Spannungen während der Abkühlung, auftrat.

Diese Arbeit schließt mit einer Fallstudie über ein supraleitendes Kabel in einer Aluminium-Elektrolyseanlage. Es wird gezeigt, dass die Energieeinsparungen für ein 500 m langes supraleitendes Kabel der Klasse 100 kA im Vergleich zu einer herkömmlichen Aluminium-Sammelschiene im Bereich von 6 GWh bis 6,5 GWh pro Jahr liegen, was in etwa dem jährlichen Energieverbrauch von 2000 2-Personen-Haushalten in Deutschland entspricht. Die Investitionskosten für das oben genannte supraleitende System sind derzeit drei- bis sechsmal höher als bei einem herkömmlichen Aluminium-Sammelschienensystem. Der große Unterschied in den Investitionskosten des supraleitenden Systems ist weitgehend auf eine große Varianz in den Supraleiterkosten zurückzuführen, die die Investitionskosten dominieren. Der Break-Even-Punkt zwischen dem supraleitenden System und dem konventionellen System für die minimalen Investitionskosten beider Systeme liegt bei einem Strompreis von ungefähr 30 €/MWh, berechnet über einen Zeitraum von 40 Jahren.

Contents

3 Introduction

The threat of climate change has captured the attention of millions of peoples and motivated politicians around the world to take measures to limit the emission of green house gases. In the last decade Germany started with the *Energiewende* a nation wide project to move its electricity production away from nuclear and fossil fuel based technologies towards renewable energies in an attempt to limit the nation's impact on global warming. In this framework, Germany aims to reduce its primary energy consumption by 20 % in 2020 and 50 % by 2050 compared to its 2008 consumption [Bun19; Bun]. A large part of the primary energy reduction will be achieved by replacing fossil fuel based technologies with renewable energies. Nonetheless, efforts to increase energy efficiency are needed in particular since the primary energy reduction in 2018 was only about 10 % compared to 2008 [App+19]. Considering the net power consumption in Germany by sectors, shows that the industry in general is the largest consumer with roughly 46 % [App+19]. About 50 % of the industries net power consumption (or 23 % of Germany's net power consumption) can be attributed to a few so called energy intensive industries such as steel, paper, glass, chemicals and metals production [Gei19]. If one aims to increase energy efficiency, then it appears to be prudent to investigate the energy savings potential in the largest consumers.

One technology to improve energy efficiency that has been discussed for decades is the loss free transport of electricity via superconducting cables [Gra97]. However, early concepts with low temperature superconductor were not competitive due to the high cooling effort. This changed with the advent of commercially available high temperature superconductors in the last two decades that have the advantage of requiring only liquid nitrogen for cooling. Consequently, several superconducting cable projects set out to demonstrate the technical feasibility of the technology with the longest cable so far being installed in Essen and being in operation for more than 5 years without any interruption [Ste+13; Her+16; Her18; Mag+11; MYR15]. The early high temperature superconducting cables focused mainly on the AC distribution grid [Yum+09; Kim+13b; Sch+12] and only recently DC cables have gathered increased attention [Wat+17; Yan+15; Zha+15].

This thesis aims to develop high direct current, high temperature superconducting cables for example for energy intensive industries. For this purpose chapter 4 reviews the basic properties of superconductors in general and high temperature superconductors in particular that are relevant for the application in the high current sector. Chapter 5 reviews the advantages of superconducting power transmission and follows up with a summary of the state of the art in regards to high current power transmission with high temperature superconductors.

In Chapter 6 a conceptual design process for high temperature superconducting cables for high direct current application is developed. The design process is used to generate a conceptual design of a 100 kA cable which is then down scaled to a 3.6 m long 35 kA cable demonstrator for first proof of concept.

Manufacturing a superconducting cable is a challenging task due to the mechanical and thermal properties of the basic superconducting wires. Therefore, the performance degradation of high temperature superconducting wires under various thermal loads is investigated in detail in Chapter 7.

Chapter 8 describes in detail the electrical and microscopic characterization of the demonstrator strands and compares them with the expected performance based on calculations. The individual characterization allows for an estimation of the current carrying capacity of the cable demonstrator which was tested in a newly designed cryostat.

To apply the design methods developed in this thesis, a case study for superconducting high current cable in an aluminum electrolysis plant is performed in Chapter 9. The study consists of a conceptual design of the cable based on the design process developed in the Chapter 6 and an economic evaluation of the superconducting cable compared to a conventional solution based on a present value calculation of the investment.

The results of this thesis are summarized in Chapter 10 and an outlook for future work is given.

4 Basics of high temperature superconductivity

At the beginning of the 20th century, Heike Kamerlingh Onnes measured the electrical resistance of various materials when cooled to very low temperatures. In 1911 he observed that the electrical resistance of mercury vanished at a temperature of 4.2 K [Kam11]. Since these materials behaved as superior conductors he called the effect superconductivity. In following years it was discovered that many materials can reach the superconducting state when cooled down to sufficiently low temperatures [Sha15, p. 17-18]. New materials were found that turned superconducting at slightly higher temperatures but it appeared that superconductivity was limited to temperatures below 30 K [Man11].

This changed in 1986 when Bednorz and Müller discovered a new type of superconductor [BM86]. Based on this discovery, soon a wide variety of materials were found that were superconducting at temperatures in the 90 K range [WU+87]. These new superconductors could not be completely described by the existing theories and are referred to as high temperature superconductors (HTS) to differentiate them from the classical, low temperature superconductors (LTS).

A hundred years after the discovery of superconductivity, new superconductors are still being found for example Magnesium diboride (MgB_2) in 2001 [Nag+01], iron-based superconductors in 2006 (since 2008 with critical temperatures above 55 K) [Kam+06; Tak+08; Ren+08], hydrogen sulphide (H_2S) in 2015 at 203 K at high pressures [Dro+15] or lanthanum hydride (LaH_{10}) in 2019 at 250 K at high pressures [Dro+19].

4.1 Critical values of superconductivity

There are three so-called critical values that limit the superconducting state

- critical temperature T_c
- critical magnetic field B_c
- critical current density j_c

The critical values are dependent on each other as it is illustrated in Figure 4.1. The critical values and their mutual dependence are inherent material properties that depend in addition on the various processes that are used to manufacture superconductors. In order to accurately describe an operation point, all three critical values need to be defined. For example critical current values are often reported at a defined temperature under self field (s.f.) conditions.

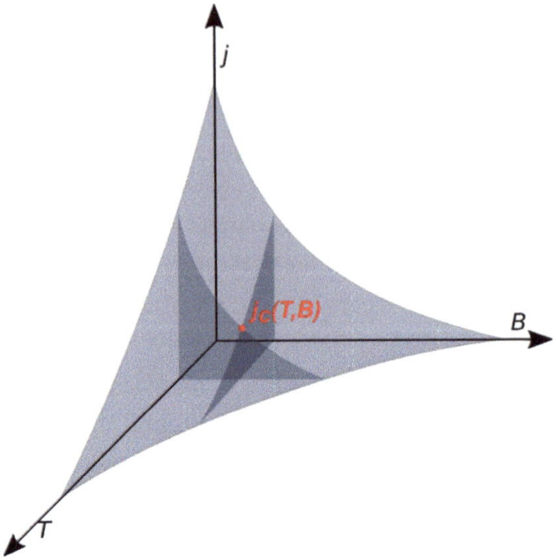

Figure 4.1: Schematic of the interdependence of the critical values for HTS. The three critical values create a surface that sets the boundary for the superconducting state.

4.1.1 Critical temperature

Figure 4.2 shows the electrical resistance of a HTS as a function of the temperature. The resistance is normalized to the linear decrease of the normal conducting phase. The transition between normal conducting state and superconducting state is usually steep. It is common practice to define the transition width as well as a critical temperature. Within the normalized values certain thresholds of the electrical resistance are defined (usually 10% and 90%) and the transition width calculated as $\Delta T_c = T_{c,90} - T_{c,10}$ [KC17, P. 1228].

4.1.2 Critical current density

In Figure 4.3 the electrical field of a HTS is schematically shown as a function of the applied current. Initially, there is no electrical field across the HTS however at a certain current an electrical field starts to develop. The critical current is defined at a certain electrical field threshold for LTS: $E_c = 0.1\ \mu\text{V/cm}$ [DKE07a; DKE07b] and for HTS: $E_c = 1\ \mu\text{V/cm}$ [DKE07c].

This current-electric field behavior can be modeled in the transition area with a power law dependence. [Gri+14a]

$$E = E_c \left(\frac{j}{j_c}\right)^n \tag{4.1}$$

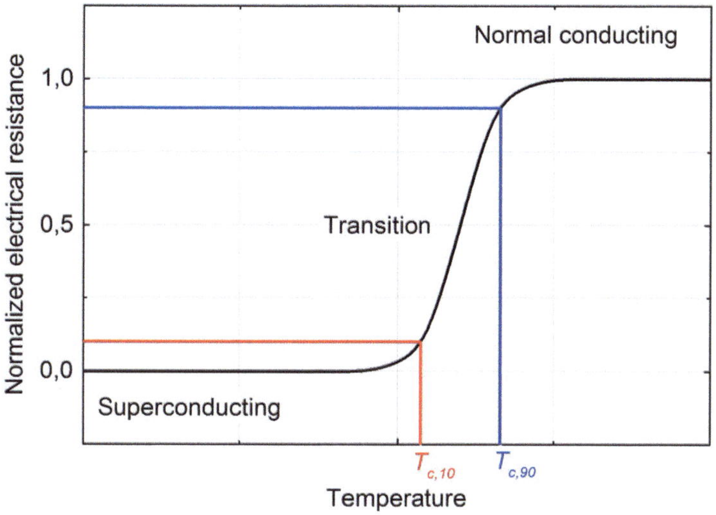

Figure 4.2: Electrical resistance of a superconductor normalized to the linear decrease of the normal conducting phase. Definition of two critical temperatures at 10 % and 90 % electrical resistance.

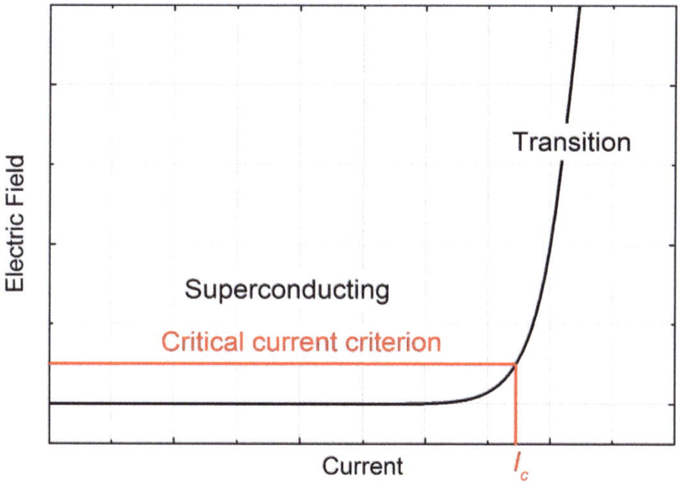

Figure 4.3: Schematic of electrical field-current behavior of a HTS. For HTS the critical current is reached when the electrical field reaches 1 µV/cm.

with E_c being the standardized, critical electrical field for example $E_c = 1\,\mu V/cm$ in case of HTS. The exponent n represents the steepness of the phase transition from superconducting to normal conducting. In this work the n value is determined between E_c and $10 \cdot E_c$ when possible.

Reports of the current carrying capacity of superconducting wires usually include either the critical current I_c of the wire, the critical current density j_c, or the engineering critical current density j_{ce}. The current densities j_c and j_{ce} differ by the cross section used to normalize the critical current. The critical current density normalizes to the superconducting cross section of the wire:

$$j_c = \frac{I_c}{A_{Superconductor}}$$

The engineering critical current density normalizes to the entire wire cross section:

$$j_{ce} = \frac{I_c}{A_{Wire}} \tag{4.2}$$

4.1.3 Critical magnetic field

Figure 4.4 shows a schematic magnetic phase diagram of cuprate HTS. There are three distinctive values: the Meißner phase limit B_{c1} , the superconducting phase limit B_{c2}, and the irreversibility field B_{irr} [Sha15; Lar+01; KC17]. The irreversibility field B_{irr} describes the maximum field to which a superconductor can transfer a current. Beyond B_{irr} the critical current density of the superconductor is zero and therefore describes the limit for application relying on loss free current transfer such as cables or magnets.

An empirical formula for estimating the irreversibility field B_{irr} is

$$B_{irr} = B_{irr}(0\,\mathrm{K})\left(1 - \frac{T}{T_c}\right)^{\alpha} \tag{4.3}$$

with being α a material characteristic fit parameter [KC17, p. 1240].

The magnetic field acting upon a superconductor can either be self-induced or externally induced. Self-induced magnetic fields occur in all superconducting devices that carry a current such as wires or cables. The self-induced magnetic field is often referred to as self-field (s.f.). Externally induced fields refer to fields generated by other devices such as magnets in the vicinity of the application and are also called background-field.

4.2 Technical superconductors

Most superconductors are not suitable for large scale applications such as cables or magnets due to toxicity, weak superconducting or mechanical properties, or expensive manufacturing. The superconductors that are used for large scale applications are often referred to as technical superconductors.

Table 4.1 summarizes some basic properties of technical superconductors. It is important to note that the properties may vary. For example, the high temperature superconductors

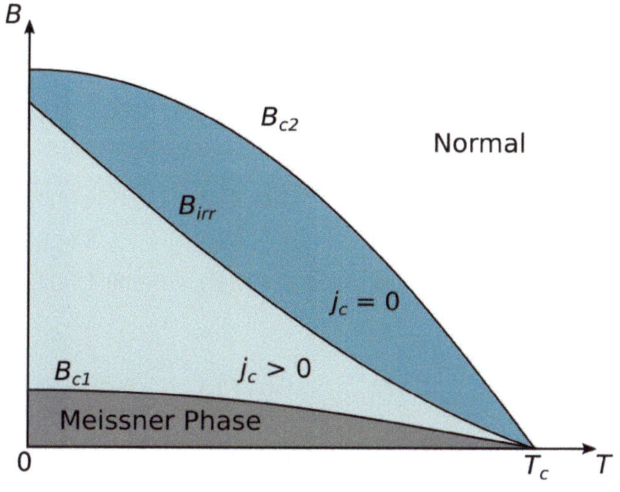

Figure 4.4: Simplified magnetic phase diagram of cuprate HTS (after [KC17, p. 1240])

show significant anisotropies in regards to their magnetic properties. Nonetheless, the shown properties allow for a basic understanding of the boundaries of these superconductors. The low irreversibility field of Bi-2223 at 77 K makes it suitable for some power cables but is already too low for high current cables as will be shown later in this work. In these scenarios *RE*BCO may be used instead, to fields up to about 7 T at 77 K. The irreversibility field of *RE*BCO shows also the limit for high field magnets operated in the liquid nitrogen region, suggesting that high field magnets with fields larger than 10 T will not be possible at these temperatures. The remaining HTS, Bi-2212, is technically a high temperature superconductor but is largely developed for high field applications in the low temperature region [Mia+05]. This is due to the fact that Bi-2212 can be manufactured in round strands contrary to the the flat tape geometry of Bi-2223 and *RE*BCO.

Magnesium diboride was discovered in 2001 and therefore is a relatively new material. Due to the wide availability of its raw materials it is considered a prospective low cost superconductor that can operated at liquid hydrogen temperatures. This could for example be applied in feeder lines towards low temperature magnets [BF17].

4.3 Second generation HTS

Rare earth barium copper oxide (*RE*BCO) represents a group of HTS that have a molecular formula of $RE_1Ba_2Cu_3O_\delta$. The oxygen content δ is in the region of $\delta = 7$ but may vary slightly depending on the targeted properties, manufacturing process or chosen rare earth. The first superconductor of this type used yttrium as a rare earth and has a critical temperature of equal to

Table 4.1: Basic properties of technical superconductors

Synonym	Chemical Formula	T_c/ K	B_{c2} / T	B_{irr} / T	Reference
NbTi	NbTi	9.5	12 (4.2 K)	10.5 (4.2 K)	[Lub83; Lar+01]
Nb_3Sn	Nb_3Sn	18	27 (4.2 K)	24 (4.2 K)	[Lar+01]
MgB_2	MgB_2	39	15 (4.2 K)	8 (4.2 K)	[Lar+01]
Bi-2212	$Bi_2Sr_2CuO_{6+\delta}$	96	>100 (4.2 K)	>100 T (4.2 K)	[Mia+05; Che+07; Jia+11]
Bi-2223	$Bi_2Sr_2Ca_2Cu_3O_{10+\delta}$	122	>10 (77 K)	~0.2 T (77 K)	[Cha+03; Lar+01]
*RE*BCO	$YBa_2Cu_3O_\delta$	93	>10 (77 K)	7 (77 K)	[Hän+07; Nak+98]

93 K [WU+87]. Yttrium can be substituted by many rare earths such as Gadolinium, Dysprosium etc. [Sha15, P. 80].

Figure 4.5 shows the crystal structure of YBCO. The superconducting current transport occurs within copper oxide planes of the crystal and is limited perpendicular to these planes thus resulting in strong anisotropic electrical current transport properties in the superconducting state [BK07, P. 100-104].

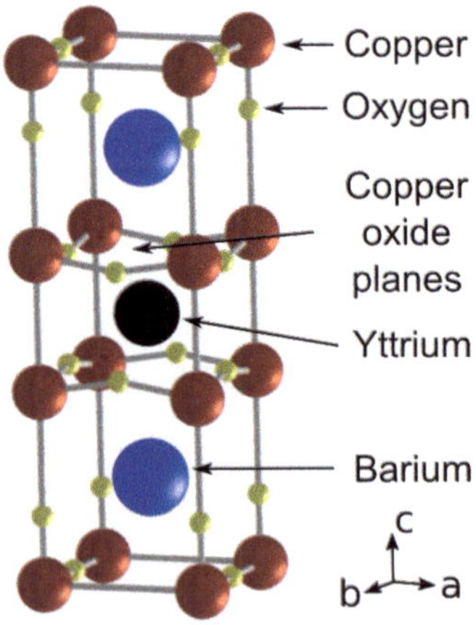

Figure 4.5: Structure of YBCO crystal . The superconducting current transfer takes place within the copper oxide planes. (after [KC17, p. 1236])

4.3.1 Structure and properties of *RE*BCO wires

In polycrystalline *RE*BCO (for example in long length wires) the superconducting grain to grain current transport is strongly dependent on the grain boundary angle and decreases exponentially with misalignment [KC17, p. 1242]. To prevent grain misalignment, *RE*BCO layers can be grown epitaxially on a previously textured layer. In the growth process of the *RE*BCO layer defects will inevitably occur and decrease the amount of well aligned grains as the layer gets thicker. Due to this, the *RE*BCO layer thickness is currently limited to a few microns. Maximizing the surface on which the *RE*BCO can be grown lead to the flat tape geometry (s. fig. 4.6) that is today the standard in long length *RE*BCO wires.

Figure 4.6 shows the structure of a *RE*BCO wire which consists of at least 5 different layers.

- Substrate material that provides mechanical stabilization for the wire but can also contain the texture (RABiTS)
- Buffer layers that act as diffusion barrier between substrate and superconductor as well as providing the texture for the *RE*BCO layer
- *RE*BCO layer which provides the superconducting current transport
- Silver layer that acts a first layer of protection but also allows the adjustment of the target oxygen content in the crystal
- In addition *RE*BCO tapes are often surrounded by an electrical and thermal stabilizer (for example copper) to protect the tape during over current events

There are three established methods to provide a texture for the *RE*BCO crystals: rolling assisted biaxially textured substrates (RABiTS), ion beam assisted deposition (IBAD) and inclined substrate deposition (ISD) [PI04]. In RABiTS the substrate is textured and the buffer layers retain the texture of the substrate. In IBAD and ISD the texture is directly deposited into one of the buffer layers.

For the deposition of the *RE*BCO layer there are also several methods employed. They can be grouped into the physical vapor deposition processes such as pulsed laser deposition (PLD), reactive co-evaporation (RCE), metal-organic chemical vapor deposition (MOCVD), and into chemical solution deposition processes such as metal-organic deposition (MOD). Detailed descriptions of the texturing and depositing technologies can be found in [[Mat+09; PI04; GPS04], [Sei15] p. 256ff.].

At present, more than a dozen companies develop *RE*BCO tapes. Table 4.2 lists the companies, the technologies employed to manufacture *RE*BCO tapes, and the critical current at 77 K, self-field normalized to the tape width. At the point of writing, several of the manufacturer were in preproduction or prototype stages. It can be seen that there is large spread in the critical

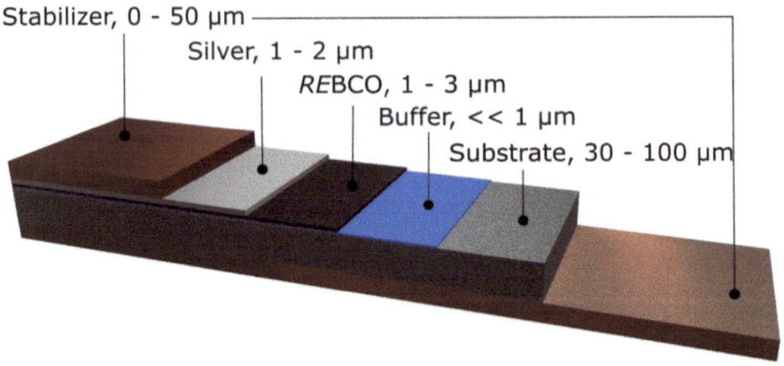

Figure 4.6: Schematic structure of *RE*BCO tapes. The thickness of the individual layer can vary depending on the manufacturing processes employed. The surrounding stabilizer material is optional but usually included in commercial tapes to act as electrical and thermal buffer in case of fault events.

currents from 60 A/cm up to 400 A/cm. This can be due to the fact that manufacturers optimize their products for different applications. For example, Bruker HTS is tailoring their tape towards extremely high field (20+ T) at very low temperatures (4.2 K) [Uso+18]. The critical current of the SuNAM tapes are optimized for the 77 K, low field region (B < 1 T). Most manufacturers can offer single piece lengths in the range of 200 to 500 m. For shorter piece length many manufacturers can offer higher critical currents than what is listed in table 4.2.

Over the past years, the top ten suppliers of *RE*BCO tapes have delivered in sum about 3000 km of tapes (in 4 mm wide equivalent). They anticipate that they will increase the supply of 4 mm equivalent tapes to about 4500 km/a until 2020. [Hol17]

4.3.2 Critical current magnetic field dependence

The critical current density of HTS depends strongly on the amplitude of the magnetic field and in general on its orientation. Figure 4.7 shows the angular critical current dependence of *RE*BCO tapes from three different manufacturers. It can be seen that the magnitude as well as the position of the peak currents varies strongly among the three manufacturers. These different behaviors can be attributed to the different manufacturing processes that are used.

As seen in Figure 4.7 a precise knowledge of the orientation of the superconductor within the magnetic field is required to accurately calculate its critical current in practical applications. Furthermore, the critical current angular dependency of the superconductor has to be known and an accurate model to reflect it has to be determined. It was shown in [Gri+14b] that a model as shown in equation 4.4 is able to reproduce measurements within a 10% error

$$j_e(B,\theta) = \frac{j_{e0}}{\left[1+\sqrt{(k_1\cdot B\cdot\cos(\theta+\alpha))^2+(B\cdot\sin(\theta+\alpha))^2}/B_k\right]^{k_2}} \tag{4.4}$$

Table 4.2: List of *RE*BCO tape manufacturers and available, normalized critical currents of selected manufacturers at 77 K, s.f.

Company	Process	Texture	I_c / width	Ref.
American Superconductor (AMSC)	MOD	RABITS	300 A/cm	[Tsu+17]
SuperPower Inc. (SPI)	MOCVD	IBAD	400 A/cm	[Haz17]
SuNAM	RCE	IBAD	400 A/cm	[Lee17]
SuperOx	PLD	IBAD	400 A/cm	[Mol17]
Fujikura	PLD	IBAD	400 A/cm	[Fuj17]
Bruker HTS	PLD	ABAD	100 A/cm	[Uso+18]
Theva	RCE	ISD	400 A/cm	[Bau17]
Superconductor Technologies (STI)	RCE	IBAD	300 A/cm	[SN16]
Shanghai Superconductor Technology (SST)	PLD	IBAD	300 A/cm	[SN16]
Deutsche Nanoschicht (DNA)	MOD	RABITS	300 A/cm	[Bäc17]
Shanghai Creative Superconductor Technologies	MOD	IBAD	170 A/cm	[Tsu+17]
Oxolutia	MOD	IBAD	60 A/cm	[Vla+18]
SWCC Showa Cable Systems	MOD	IBAD	350 A/cm	[Tsu+17]
MetOx	MOCVD	RABITS	300 A/cm	[Met]
Samri	MOCVD	IBAD	300 A/cm	[Sam]

with B_k, k_1 and k_2 being model parameters , j_{e0} the critical current density in self field , B the magnetic field magnitude , θ the field angle, α the displacement of the critical current peaks from 0°. Based on the procedure described in [Gri+14b] the model parameters were determined from ITEP internal data for six manufacturers (see Table 4.3).

4.3.3 Critical current temperature dependence

Figure 4.8 displays the critical current per centimeter tape width of commercial *RE*BCO tapes at different temperatures for various manufacturers. The temperature dependence follows a roughly linear behavior for large temperature ranges. When comparing temperature dependencies of several tapes or manufacturers it is common practice to normalize the critical current values to its 77 K, s.f. value. These normalized values are often called lift factors (LF).

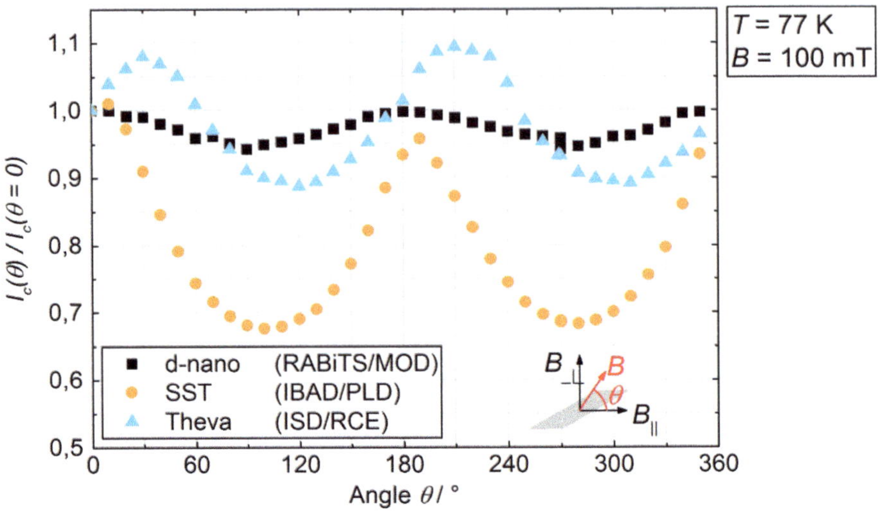

Figure 4.7: Angular critical current dependence of *RE*BCO tapes from deutsche Nanoschicht, Shanghai Superconductor Technology and Theva at 77 K and 100 mT background field. (Data published in [Sch17])

Table 4.3: Angular dependencies parameters at 77 K, s.f. conditions

Manufacturer	Average Error / %	B_k	k_1	k_2	α	Data range / mT
SST	3.5	0.15	0.35	0.95	10	50-600
DNA	4.4	0.6	0.85	1.95	0	50-600
Theva	2.5	0.25	0.7	1.05	25	50-600
SuperOx	5.8	0.25	0.45	1.45	10	25-600
SPI	4.5	0.15	0.2	0.9	3	50-600
SuNAM	6.8	2.65	0.45	10	0	50-200

$$LF(T) = \frac{I_c(T)}{I_c(77K)}$$

Figure 4.9 shows that the temperature dependence is very similar among the manufacturers within the liquid nitrogen region (65-77 K). At 65 K the smallest increase is observed with SuNAM tapes with a lift factor of 1.86 while the highest increase is shown in the SPI tapes with a lift factor of 2.37.

In summary it can be noted that the critical current temperature dependence of *RE*BCO tapes is significant within the liquid nitrogen region with lift factors of up to about 2.5. Variation in the temperature dependence among different suppliers can be observed.

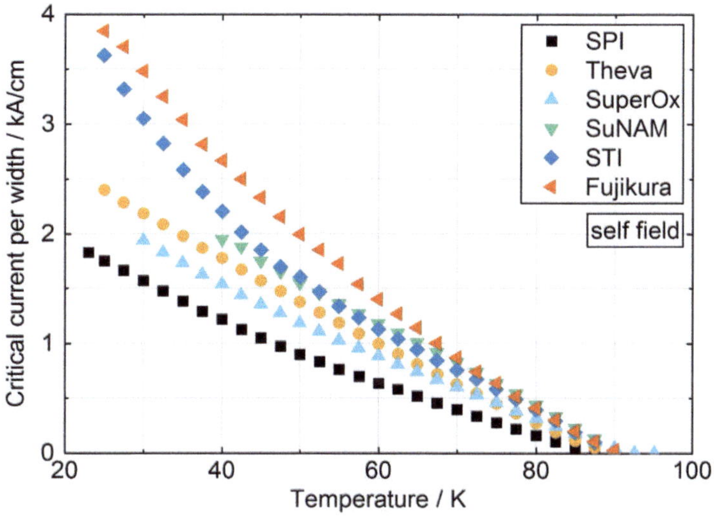

Figure 4.8: Critical current - temperature dependence of *RE*BCO tapes in self-field conditions (based on data published in [SN16])

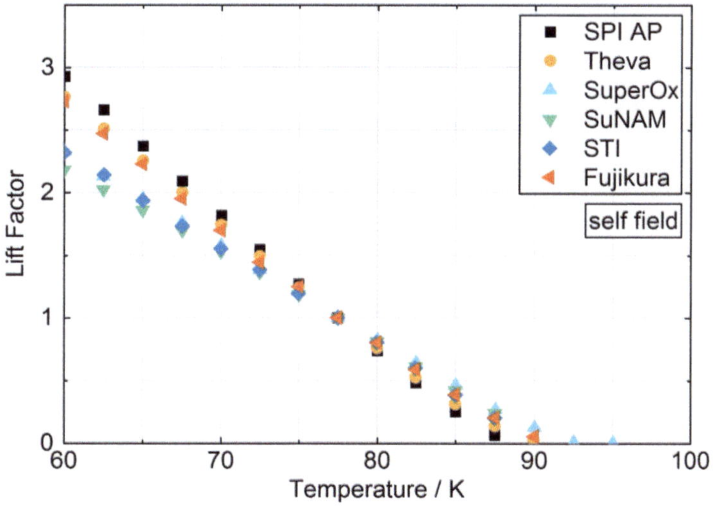

Figure 4.9: Lift Factors of *RE*BCO tapes in self-field conditions (based on data published in [SN16])

4.3.4 Critical current strain dependence

The critical current of a *RE*BCO tape strongly depends on its strain. In [BMS15] the electro-mechanical properties of REBCO tapes from several manufacturers were studied. The results of that study are summarized in Table 4.4. The irreversible strain limit ε_{100} is the strain value after which the critical current did not return to its original value when the strain was relaxed. It is therefore the threshold for permanent mechanical damage to the conductor. A reversible reduction of the critical current is observed before the damage becomes permanent. For the studied superconductors in [BMS15] this reversible reduction started at around 0.4 % strain for all manufacturers.

Table 4.4: Summarized electro-mechanical properties measured in [BMS15] at 77 K in self field.

Manufacturer	Irreversible strain limit, ε_{100}/ %	Irreversible stress limit, σ_{100}/ MPa
Bruker HTS	0.70-0.72	660-670
Fujikura	0.56-0.58	690-700
SuNAM	0.66-0.68	740-750
SuperOx	0.47-0.49	740-760
SPI	0.66-0.69	800-820

5 State of the art of *RE*BCO high current transmission

5.1 High current applications

Voltage levels such as 'high voltage' are defined by standards of the IEEE, VDE or other standardization organizations. This is not the case for currents. The term 'high current' can mean different currents depending on the context of the work. For the purpose of this work 'high current' is defined as currents above 10 kilo amperes.

The advantages of superconductors in general and *RE*BCO in particular for direct current transmission have been recognized in several pilot and demonstrator projects around the world (see Section 5.3). In the following it will be discussed how these advantages translate into the high current regime.

Advantages

- *High current density*
  Conventional copper bus bar systems have, according to DIN 43671 [Ger00], a current density in the area of 1 A/mm^2 and even less if the required distance between individual bars is considered. Single *RE*BCO tapes have an engineering current density of 400 A/mm^2 in 77 K, s.f. and more (see Table 4.2). Of course these current densities can not be achieved in a cable system due to the cooling requirements and the critical current magnetic field dependence. However, it shows how strong the current carrying capacity can be upscaled once a coolant supply is in place. This can be used to upgrade the capacity of existing transfer lines by substituting them with superconducting ones (retrofitting) or create significant space and weight saving [Mic+15] for example in data centers or industrial electrolysis facilities.

- *Zero resistance transmission*
  Superconducting transmission systems can achieve lower system losses than conventional bus bar systems in the high current regime [Mic+15]. This advantage increases for higher currents and operational times as the ohmic losses scale with the power of two to the current. Industrial electrolysis is one extreme application with currents up to 600 kA while also having operational times of 23 h per day or more [KD14; FM06; Gar04; OBH07; GE ; Mor15]. Chapter 9 describes a case study for a *RE*BCO power transmission system in industrial electrolysis.

The lower system losses are also represented in a lower voltage drop along the a transmission line. This is an attractive characteristic for example in feeder lines for public transportation. In order to keep the voltage drop within acceptable bounds, the Japan Railway requires in Tokyo, Japan, substations every 2-3 km. A superconducting solution could significantly reduce the numbers of substations [Tom+12]. It could also allow new layouts in for example data centers utilizing a facility wide low voltage network [Min+09; PKA07].

- *High critical temperature*
  The high critical temperature of *RE*BCO compared to for example MgB_2 allows simple, reliable LN_2 cooling systems utilizing redundancies that exist in large scale nitrogen liquefiers [HKI14; Sau17]. Furthermore it also simplifies the design of the cryostat as a simple two-walled vacuum isolated cryostat is usually sufficient.

5.2 High current conductor concepts

One of the challenges of high current transfer with HTS is that the current of the individual tapes is limited to a few hundreds amperes. Depending on the desired current, tens or hundreds of tapes need to be combined to achieve the operational currents. This is further complicated by the flat HTS tape geometry and strain dependence of the critical current which makes it difficult to directly employ cabling techniques from for example LTS or conventional power cables. Currently there are three distinctly different cabling approaches which will be explained in detail.

In order to avoid confusion within this work the following terms are defined as

- Conductor: General term for a superconducting element transferring electrical energy of any size
- Wire: Superconducting base element for strands and cables, usually *RE*BCO tapes within this work, also called coated conductors in the literature
- Strand: Arrangement of several wires
- Cable: Arrangement of several wires or strands into a functional unit including a cryostat
- Conventional: Element with an ohmic resistance

5.2.1 Co-axial winding concept

Figure 5.1 illustrates the concept of strands based on wound *RE*BCO tapes. The tapes are wound

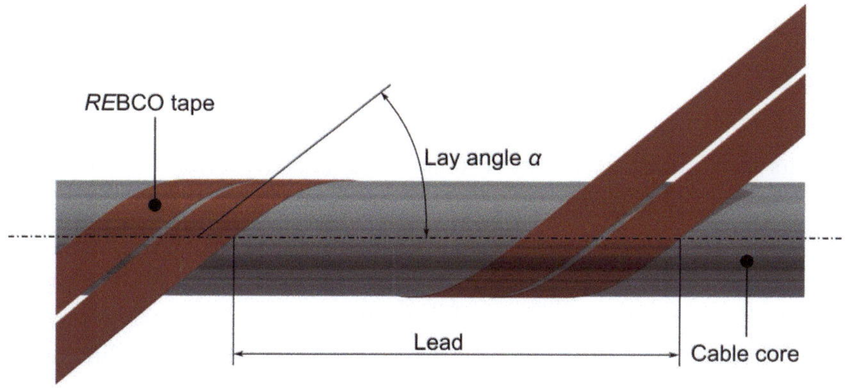

Figure 5.1: Schematic of *RE*BCO co-axial winding concept. *RE*BCO tapes are wound around a cable core under the lay angle α and a certain lead.

around a central core with a diameter d and a lead p utilizing the good bend ability of the *RE*BCO tapes around their flat side. The lay angle α is calculated by

$$\alpha = \arctan\left(\frac{\pi d}{p}\right) \tag{5.1}$$

There are two variants of this concept

- large core and small lay angle e.g. d = 35 mm, $\alpha = 15°$ in [Kim+13c]
- small core and high lay angle e.g. d = 5 mm, $\alpha = 30°$ in [vGH12]

The first variant has been used in several AC and DC cable projects around the world [Yum+09; Mas+05; Kim+13b; Cho+11; Lee+11; Kim+13c; Yam+15; Yan+15]. In this case the superconductors are roughly aligned with the cable self field which can be beneficial to minimize the impact of the angular critical current dependency of the superconductor (s. Fig. 4.7).

The second variant is currently developed as a high current strand by Advanced Conductor Technologies (ACT) under the brand name Conductor on Round Core (CORC). To achieve high currents multiple layers of superconductors are wound onto each other under alternating lay angles into a single CORC strand. Due to the high lay angle a CORC strand offers very small bending radii for example 5 cm with a minimum of degradation [van+13; van14] and has an isotropic field dependence due to alternating the lay angle [Bar+15]. These two benefits make the CORC variant an attractive option for magnets. The CORC strand is also one of the few *RE*BCO high current strand concepts that are commercially available. In 2016, approximately 120 m of CORC strand were produced of which about 70 m were commercial orders [Wei17].

An inherent disadvantage of wound concepts is that for a meter of the conductor more than one meter *RE*BCO tape is required. As can be see from Figure 5.1 it scales with the lay angle α as

$$L_{REBCO} = \frac{L_{cable}}{\cos\alpha} \tag{5.2}$$

Table 5.1 lists some exemplary results of co-axial wound conductors. The 500 m co-axial cable is a fully functional *RE*BCO demonstrator cable that operates on Jeju Island, Korea. It features two *RE*BCO tapes layers which have a lead of 290 mm and 320 mm on core radii of 10.7 mm and 11.4 mm respectively. The cable is designed to operate at 3125 A and 80 kV with a maximum operating temperatures of 72 K. It was installed in 2015 to connect two substations and is so far the only long length DC cable that uses *RE*BCO tapes as superconductors [Sim+13; Yan+15].

The 6 around 1 CORC follows the classical cabling structure with a central tube for the coolant and all encased in an aluminum jacket. It was designed for large high field magnets (~20 T) such as fusion or detector magnets. Each of CORC strand contains 38 4 mm wide *RE*BCO tapes spread over 12 layers. The tapes are wound around a 5.2 mm core resulting in an outer diameter of 7,6 mm of the CORC strand. [Mul+17; Mul+16b; Mul+16a; Mul+16c]

The 2 phase CORC system was designed to be used on US Air Force on-board applications that require currents of up to 18.5 kA at 270 V. The two phases are arranged in a concentric manner. The inner phase consists of 10 layers of *RE*BCO tapes with radii between 2.77 mm and 4.12 mm and lead between 30 mm and 68 mm. The outer phase consists of 7 layer of *RE*BCO tapes with radii between 4.2 mm and 5.1 mm and lead between 52 mm and 69 mm. The system was rated to carry 20 kA at 55 K per phase and achieved 7.5 kA at 76 K self-field [vGH12].

Table 5.1: Experimental results of CORC and co-axial winding concept cables using *RE*BCO tapes

Type	Critical current	Length	References
Co-axial winding cable	4103 A 72 K s.f.	500 m	[Sim+13]
CORC: Two Phase	7561 A 76 K, s.f.	~1 m	[vGH12]
CORC: 6 Around 1	13000 A, 77 K s.f.	0.8 m	[Mul+17]

5.2.2 Roebel concept

A Roebel strand consists of several *RE*BCO tapes from which a portion of the tape has been punched out. These meandering tapes can be assembled into a cable as is shown in Figure 5.2.

The advantages of the Roebel concept are transposition of the tapes which results in relatively low AC losses [Gol+14], good bend ability around the flat site of cable [Ott+16] and possibly favorable alignment to an external magnetic field.

However due to the punching process material loss is unavoidable which can be a challenge considering the high cost of HTS. Applications where the conductor experiences high stresses

Figure 5.2: Schematic of Roebel cable consisting of 5 tapes [Bar13]

are also challenging for the Roebel cable as there can be degradation at the edges [Bay+15]. A way to mitigate this is to add for example resins to improve mechanical stabilization [Ott+15]. Scaling up of the Roebel strand is also limited due to the fact only a certain amount of tapes fit within on twist pitch [Gol+14].

The Roebel concept is developed by the KIT and General Cable Superconductors (GCS). Table 5.2 lists some basic properties of Roebel strands prepared in the past.

Table 5.2: Basic results of Roebel strands produced in the past

Group	I_c 77 K,s.f.	Length	Number of tapes	Punched tape width	Ref.
KIT	447 A	1 m	10	1.8 mm	[Gol+14]
KIT	2628 A	1.1 m	45	5 mm	[Gol+14]
KIT	936 A	5 m	10	5.5 mm	[Gol+14]
KIT	1411 A	1.1 m	15	5.5 mm	[Ott+16]
GCS	309 A	0.54 m	9	2 mm	[Gol+14]
GCS	1100 A	5 m	15	5 mm	[Gol+14]
GCS	1420 A	21 m	15	5 mm	[Gol+14]

5.2.3 Stack concepts

In stack concepts, individual *RE*BCO tapes are stacked on top of each other. There are two main variations of this approach. In the first approach tapes are stacked until the design critical current is achieved. In the second approach a certain amount of tapes is stacked and then embedded into casing or matrix to create a round strand. Multiple of these round strands can than be assembled to achieve higher critical currents.

The first approach has the advantage of being easily scalable to very high currents as was demonstrated at the National Institute for Fusion Science (NIFS) in Japan. Where a demonstrator was build that achieved 100 kA at 20 K and 5 T background field. The demonstrator uses three directly adjacent stacks of 18 10 mm wide *RE*BCO tapes totaling 54 tapes each having a critical current of about 600 A at 77 K, s.f. The conductor was build to demonstrate the feasibility of a stacked tape conductor approach in a helical fusion reactor [Ter17; Ter+15; Ter+14; Ito+14].

A joint development of Vision Electrics Super Conductors (VESC) and Karlsruhe Institute of Technology (KIT) also used the first approach to develop a commercial stacked conductor for industrial applications. The goal was to develop stiff conductor elements that can be quickly assembled on site and easily connected via low resistive joints. The concept uses two stacks of *RE*BCO tapes with each tape being separated by a copper spacer to reduce the impact of the magnetic field. The joints developed in the project achieved contact resistance below 1 nΩ between to stacks. A prototype with two stacks of 23 *RE*BCO tapes reached 20 kA at 77 K [Els17].

The second approach is followed by the Massachusetts Institute of Technology (MIT) that developed a stack concept where the conductor is twisted around its central axis. The twisting helps to decrease heat dissipation in large scale magnets for example in fusion devices due to decreasing AC losses. Grooved copper rods have been investigated in which the stack can be inserted to create a round strand [Tak+12]. The stacks are usually square therefore resulting in 40 to 60 tapes for 4 and 6 mm wide tapes assuming a tape thickness of 100 µm [Tak+17].

The Swiss Plasma Center (SPC) also follows the second approach by stacking and twisting *RE*BCO tapes. Additionally the *RE*BCO tapes are soldered into two grooved copper half's thus creating a round strand. Twenty of these round strands each containing 16 4 mm wide tapes were assembled into a demonstrator conductor for a future fusion device. In total, two conceptually identical demonstrators were build with *RE*BCO tapes from different manufacturers, one was supplied by SuperPower Inc and the other by SuperOx. Both demonstrators achieved 60 kA at 5 K and 12 T background field. [Ugl+15; Byk+16]

Another concept following the second approach is the Cross Conductor developed by KIT which will be examined in detail in section 5.2.4.

In Italy at 'Agenzia nazionale per le nuove tecnologie, l'energia e lo sviluppo economico sostenibile' (ENEA) a concept is developed with focus on industrial scale production. The concept is to use an extruded aluminum profile with 5 to 6 grooves into which 20 to 30 unsoldered 4 mm wide *RE*BCO tapes are stacked. The grooves of the aluminum profile follow a helical profile with a lead of about 1.6 m. The diameter of the aluminum profile is about 20 mm. To the authors knowledge no sample has been measured where all grooves were fully stacked with *RE*BCO tapes. The expected current is 10 kA at 77 K, s.f. conditions. [Cel+14; Aug+15]

Table 5.3 summarizes basic properties of samples following the stack approach.

Table 5.3: Experimental results of stack concepts

Group	Type	Current	Intended Application	Ref.
MIT	Strand	6 kA 4.2 K 17 T	High field magnets	[Tak+17]
SPC	Strand	1.6 kA, 77 K s.f.	Future Fusion Reactor	[Ugl+15]
SPC	Demonstrator	60 kA 5 K 12 T	Future Fusion Reactor	[Ugl+15]
NIFS	Demonstrator	100 kA 20 K, 5 T	Helical Fusion Reactor	[Ter17]
ENEA	Expected Prototype Performance	10 kA 77 K s.f.	Fusion Applications	[Aug+15]
KIT, VESC	Subscale Prototype	20 kA 77K s.f.	Industrial Applications	[Els17]

5.2.4 Cross Conductor

In the Cross Conductor (CroCo) concept individual tapes are soldered together and embedded into a solder matrix. The concept is illustrated in Figure 5.3. By utilizing two different tape widths, the HTS CroCo aims to maximize the HTS cross section within the round matrix cross section. In an additional process the round strand can then be jacketed by for example a copper tube to improve thermal and electrical stability in case of a fault current event.

Currently there are three different sizes of HTS CroCo under development (s. Tab. 5.4). The largest currently developed CroCo uses *RE*BCO tapes with widths of 6 mm and 4 mm which typically achieves currents in the range of 3 kA at 77 K, s.f.. By reducing the size of the CroCo it is possible to reduce the minimum bending radius.

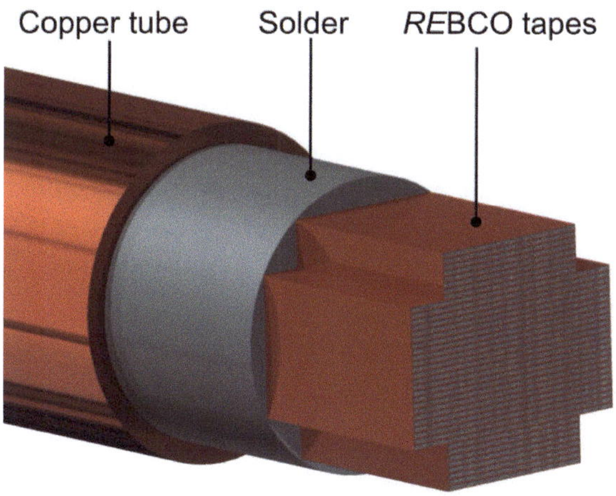

Figure 5.3: Schematic of HTS CroCo using *RE*BCO tapes of two different widths in a round solder matrix with an optional copper tube

Table 5.4: Basic composition of HTS CroCos of various sizes.

	6/4 CroCo	4/2 CroCo	3/2 CroCo
Number of tapes	22 x 6 mm	18 x 4 mm	18 x 3 mm
	10 x 4 mm	18 x 2 mm	10 x 2 mm
Ideal solder matrix diameter	7,2 mm	4,8 mm	3,6 mm
I_c(77 K, s.f.)	3200 A	2000 A	1500 A
Min. bending radius	60 cm	40 cm	30 cm

When lower current densities are possible, additional stabilizer material instead of *RE*BCO tapes, can be included into the stack. This has the advantage that it increases the distance between neighboring *RE*BCO tapes and therefore reduces the self field reduction of the critical current (Sec. 4.3.2). Additionally, the stabilizer materials helps with the current transfer into the superconducting layers. Furthermore additional stabilizer material acts as a shunt and thermal buffer to protect the superconductors from damage during a fault current event.

5.2.5 Concept comparison

In the following a conceptual comparison based on bending properties, joint and termination techniques, and tape usage is performed.

Bending

Bending is one of the challenges in *RE*BCO conductor designs due to the flat tape geometry which can be easily bend out-of-plane but hard in-plane [van+10; Ott+16]. The minimum bending radius is used as a criteria for the bending properties. It describes the radius below the conductor experiences a certain reduction in critical current (95 % I_c compared to unbend is often used). In general the minimum bending radius scales with the size of the conductor: The larger a conductor is the higher the minimum bending radius is required.

The co-axial winding concept solves the directional dependency of the bending radius. With the variation of the lay angle and lead, the minimum bending radius can be adjusted to the applications needs. The CORC strands manage minimum bending radii in the order of 5 cm for CORC strands with a diameter between 5 mm and 8 mm. Smaller CORC strands with diameter between 2.5 mm and 4.5 mm can have bending radii in the order of 2.5 cm. [Wei17]. The minimum bending radius of the full scale cable installed on Jeju Island (see Table5.1) is 1,85 m which is sufficient for transportation in drums [Yan+15].

The Roebel concept retains the good out-of-plane bend ability of the tapes with minimum bending radii in the area of 2 cm [Ott+16]. In-plane bend ability is still being investigated but likely not improved compared to tapes.

The stacked concepts usually have a minimum bending radii in the range of a few decimeters [Tak+13; Byk+15] as can be seen in Table 5.4 for the CroCo. In particular bending can be challenging for the twisted variants of stacked conductors as the conductor will be forced to bend around in-plane.

Joints and terminations

Superconducting power transmission lines longer than a few hundred meters require the use of joints between the superconducting elements. Resistance free joints have been achieved for *RE*BCO tapes [Par+14] however the process is at the moment not suitable for application on industrial sites. Therefore each joints currently introduces a resistance into the superconducting system which usually should be kept to a minimum. A challenge for low resistive joints is the layered structure of the *RE*BCO tapes with the relatively high resistive substrates. Therefore, low resistive joints can usually achieved only with access to the superconducting layer. Several methods have been investigated in the past including soldering and welding.

There are several joint concept for stacked concepts that range in their resistance from 1 to 100 nΩ [WFP17; TCM14; Els17].

For the Roebel cable joints with resistances in the range of 10 to 20 nΩ have been realized in the laboratory [Mur+18].

Joints for co-axial conductors often face the challenge of difficult access to the *RE*BCO layers placed below several other layers of *RE*BCO tape. For relatively few layers, practical industrial solutions have been demonstrated [Sch+12; Muk+09]. For a 10 layer CORC this problem has been solved by cutting away the layers in a staggered manner. With that technique a termination resistance of about 30 nΩ was achieved [Mul+15].

Tape usage

One of the challenges of *RE*BCO based power transmission is the cost of the superconductor. Therefore, it is important to analyze how efficient the various concepts use the base material. In the past, the comparison was often driven by how well the *RE*BCO tapes were aligned within a magnetic self field due to the strong anisotropic critical current magnetic field relation. In recent years a lot of progress has been made to create more isotropic tapes (see Figure 4.7). Shifting this conductor design challenge towards a manufacturer selection problem, in particular for low field applications such as power cables.

To illustrate this, critical current calculations were performed for three strand concepts and three different critical current magnetic field dependencies valid for materials from various manufacturers. For a useful comparison the dimensions and initial amount of tapes need be similar across all concepts. The outer diameter was therefore limited to about 7.6 mm. The amount of *RE*BCO tapes is defined by the sum of the *RE*BCO tapes widths and was set to 156 mm unless otherwise noted. The critical current was 30 A per mm width.

For the CORC this manifested in 39 4 mm wide tapes that were distributed over 13 layers with a layer thickness of 100 μm, similar to a CORC described in [Mul+16c]. The stack concepts were represented by a CroCo with 22 6 mm tapes and 6 4 mm tapes with 215 μm between the tapes. For the Roebel strand 40 4 mm wide tapes were used totaling 160 mm width with 320 μm distance between the tapes. The calculations were performed with the models described in section 6.3.

The results are summarized in Table 5.5. Noticeable is that the Roebel concept results in the lowest critical current with about 50 % of the critical current of the other concepts. This is not surprising since the punching process reduces the tape cross section by about 50 %.

Table 5.5: Critical currents of three strand concepts for three different critical current magnetic field dependencies. Initially all tapes have the same amount of tapes. The strong reduction of the Roebel concepts is due to the reduction of tape width during the punching process.

Manufacturer	CroCo	CORC	Roebel
SST	3406 A	3762 A	1747 A
DNA	3236 A	3283 A	1642 A
Theva	3181 A	3281 A	1627 A

The CORC strand achieves the overall highest critical current in all cases. Compared to the CroCo the critical current of CORC strand is about 10% higher for tapes with strong anisotropic angular critical current dependency (SST) but only between 1% and 3% higher for more isotropic tapes.

In addition to the critical current, one needs also to consider how much tape is necessary for one unit length of strand for an accurate estimation of the tape usage. This can be expressed in a factor such as L_{tape} / L_{strand}. For the CroCo this factor is 1 as one unit length of tape results in one unit length of strand. This is not true for the Roebel strand and the CORC strand. In the Roebel strand it depends on the transposition and the amount of tapes while the CORC strand it depends on the lay angle. According to [Bar+15] the lay angle in CORC varies between 30° and 60° which translate into a L_{tape} / L_{strand} factors between 1.15 and 2.

Summary

In this summary three strand concepts are being compared on a scale of 3 (+, 0, -) in the previously described categories.

In terms of bending, the winding concepts (CORC) have the clear advantage over all other concepts due to solving the directional dependence of the *RE*BCO tape and realizing very tight bending radii if necessary. The Roebel concepts also manages small bending radii out-of-plane but in-plane bending remains to be investigated [Ott+16]. Stacks have the poorest bending properties of the three concepts.

Joints and terminations have been developed the furthest in stack concepts with multiple suggestions on how to realizes practical low resistive joints. For the Roebel concepts promising joint concepts have also been realized. Practical low resistive joints remain a challenge in winding concepts with multiple superconducting layers.

The Roebel cable has the highest tape consumption in regards to achieved critical current per meter strand of the three concepts which is largely due to the loss of superconductor in the punching process. The winding concepts can profit from being well aligned with the magnetic field thus achieving higher critical currents than a stack for certain tape properties. However, the necessary winding around the core negates the advantage in critical current by higher tape consumption per meter strand compared to a stack. If the critical current between stack and winding concept is equal than the advantage in tape usage has to be granted to the stack concept.

The ratings are summarized in Table 5.6. In conclusion it can be noted that the Roebel concept is not suitable for bulk power transmission due to its poor tape usage. The winding and stack concepts are equal in their overall rating with the distinction in the bending properties.

Table 5.6: Rating of three strand concept based on their bending, joints and tape usage.

	Bending	Joints	Tape usage
Winding	+	0	0
Roebel	±	+	-
Stacks	-	+	+

5.3 Summary of HTS DC cable projects

Table 5.7 gives an overview of existing DC power transmission projects of significant length. It can be seen that the majority of the cables still utilizes BSCCO as superconductors. Only in recent years two projects were realized with *RE*BCO. One explanation might be the relatively long project time of such long length cables and the present availability of large amounts of HTS.

The highest currents of 20 kA so far were realized with two 20 m MgB_2 cables operated at 24 K at CERN which were part of investigations for LHC High Luminosity upgrade at CERN [BF17]. The goal of the investigation was to determine if MgB_2 cables could be used as high current feeders for the magnet systems installed at CERN.

A *RE*BCO stack concept developed by Vision Electrics Super Conductors (VESC) and KIT also achieved 20 kA at 77 K in a subscale test. The concept was unique as it features multiple joints with resistances below 1 nΩ. Full scale elements with a total length of 25 m were installed in a chlorine electrolysis at BASF in Ludwigshafen and are being tested.

In the 10-kA range a 360 m long BSCCO based cable has been in operation since 2012 in China as a connection between a power plant and an aluminum electrolysis plant. Similar currents are planned for a high-voltage MgB_2 cable demonstrator which is developed by a consortium including Nexans, KIT and CERN for bulk power transmission within the *Best Path* framework of the European Commission. The goal is to build a full scale demonstrator cable that validate the usability of MgB_2 in high power electricity transfer as well as study the economic viability of such a cable and its environmental impact.

Table 5.7: Superconducting DC power transmission projects around the world

Year	Length / m	Current / A	Voltage / kV	Superconductor	Location	Application	References
2006	20	2200	20	BSCCO	Japan	Test cable	[Ham+11]
2010	200	2000	10	BSCCO	Japan	Test cable	[Sun+13]
2012	360	10000	1,3	BSCCO	China	Aluminium electrolysis	[Zha+15]
2014	20	20000		MgB_2	Switzerland	Test cable	[BF17]
2015	500	3250	80	*RE*BCO	Korea	Substation link	[Yan+15; Sim+13]
2015	500	5000	20	BSCCO	Japan	Link PV farm - Datacenter	[Chi+16]
2016	1000	2500	20	BSCCO	Japan	Test cable	[Chi+16]
2017	25	20000	1	*RE*BCO	Germany	Chlorine electrolysis	[Els17]
2018+	2500	2500	20	BSCCO	Russia	Substation link	[Syt+15]
2018+	20	10000	320	MgB_2	Europe	Test cable	[Bru17]

6 Conceptual design of *RE*BCO DC cables

Since *RE*BCO tapes became commercially available there have been a number of HTS cable projects. They all share certain basic components that are required for HTS cables.

1. HTS section for current transfer
2. Cryostat to reduce heat leakage into the cable
3. Coolant flowing along the cable to cool to the superconducting state and remove excess heat
4. Dielectric to separate the cable from ground potential
5. Stabilizer to protect the superconductor during fault events

In the following sections equations that allow to conceptually design these core components are summarized.

6.1 Temperature and pressure profile

The purpose of the coolant is to remove heat that has leaked through the cryostat or was generated within due to dissipation. The overarching formulas describing the coolant flow can be derived from the balances of mass, momentum and energy. In case of a round cable the radial dimensions are negligible with respect to the axial dimensions. This allows us to simplify the equations to a 1D problem. When considering an incompressible fluid ($\rho = const.$) in steady state operation $\left(\frac{d}{dt} = 0\right)$ the balances are

$$\dot{m} = const. \tag{6.1}$$

$$\frac{dp}{dx} = \frac{\zeta \cdot v^2 \cdot \rho_{LN2}}{D_h \cdot 2} \tag{6.2}$$

$$\frac{dT}{dx} = \frac{1}{c_p}\left(\frac{\zeta \cdot v^2}{D_h \cdot 2} + \frac{q}{\dot{m}}\right) \tag{6.3}$$

with ζ the Darcy friction factor, v velocity, ρ density, D_h hydraulic diameter, $\dot{m}$ mass flow, c_p specific heat capacity and q heat load per unit length. A complete derivation of the formulas is noted in [Tre06]. The heat load per unit length q contains the heat leakage through the cryostat q_{cryo} as well as electromagnetic losses within the superconductor q_{SC} (see section 6.5).

$$q = q_{cryo} + q_{SC} \tag{6.4}$$

Equation 6.2 utilizes the Darcy friction factor ζ to describe the pressure drop due to friction within the cable. They are either determined experimentally or estimated with empirical formulas which depend on the Reynolds number

$$Re = \frac{v \cdot \rho \cdot D_h}{\mu} \tag{6.5}$$

with μ being the dynamic viscosity.

If a constant specific heat $c_p = const.$ along the cable length is assumed than the pressure drop and temperature increase can be written as

$$\Delta p = \frac{\zeta \cdot v^2 \cdot \rho \cdot L}{D_h \cdot 2} \tag{6.6}$$

$$\Delta T = \frac{\zeta \cdot v^2 \cdot L}{2 \cdot D_h \cdot c_p} + \frac{q \cdot L}{\dot{m} \cdot c_p} \tag{6.7}$$

with L as the cable length. Equation 6.6 is also known as the Darcy–Weisbach Equation.

The coolant channel usually has the geometry of a round central channel or ring type channel. The hydraulic diameter for a round central channel is the diameter of the channel $D_h = D$. For ring type channels it is calculated with the inner and outer diameter of the ring $D_h = D_{outer} - D_{inner}$.

The model described in this section represents a simple analysis of the balances of mass, momentum and energy and neglects the thermal conduction in radial direction entirely. For the purpose of this conceptual design it will be assumed that all temperatures in radial direction are equal to the liquid nitrogen temperature.

Liquid nitrogen cooling

Table 6.1 lists the fixed point properties of nitrogen [Spa00]. Figure 6.1 shows the phase diagram of liquid nitrogen and illustrates the operational range of *RE*BCO cable applications. Most HTS cables utilize sub-cooled LN_2 in order to increase the critical current and therefore the current carrying capacity of the cables (Sec. 4.3). Nitrogen can be liquid until the triple point at 63.151 K. With a safety margin against freezing and boiling, the operational temperature range of *RE*BCO cables is usually between 65 K and 77 K. Larger temperature differences are possible but one needs to consider that the critical current of *RE*BCO is roughly reduced by half from 65 K to 77 K (Fig. 4.8). In addition, a temperature margin for potential fault scenarios needs to be considered.

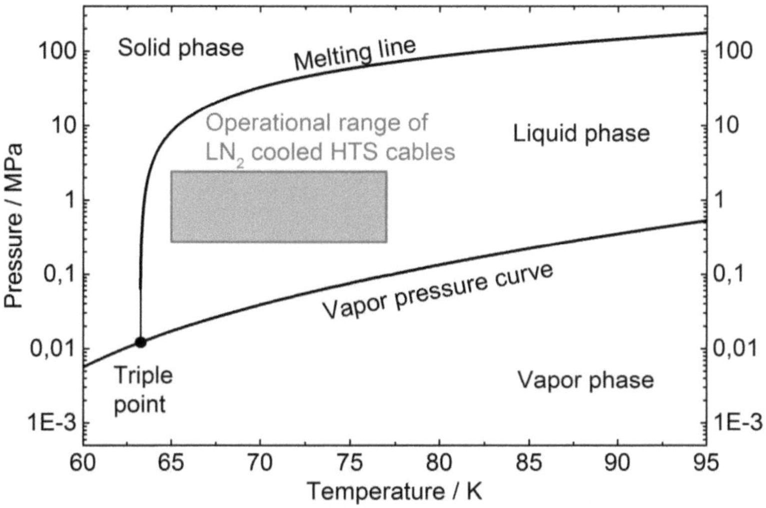

Figure 6.1: Liquid nitrogen phase diagram with a schematic of the operational range of LN-cooled HTS cables

Table 6.1: Fixed point properties of nitrogen

	Temperature (K)	Pressure (MPa)	Density (mol/dm^3)
Critical point	126.192	3.34	11.184
Triple point	63.151	0.013	

Liquid nitrogen alternatives

There are alternatives to achieve lower temperatures than 63 K. Table 6.2 lists fixed point properties of possible alternatives to nitrogen. Liquid helium is a commonly used cryogenic for LTS but the cost of 10 - 20 €/l makes it unattractive for power transmission applications. Instead gaseous helium is used in closed confines for example on ships, when the risk of suffocation also prohibits the use of liquid nitrogen [FKG07; Kep+11].

The use of liquid hydrogen is investigated as a method for combined transport of chemical and electrical energy with MgB_2 superconductors [Kos+12; Kos+15; Vys+13; Vys+15; Shi17]. However, the highly flammable nature of hydrogen may limit the use of liquid hydrogen.

Neon is usually not considered as cryogenic cooling fluid due to its lower thermal conductivity and higher price compared to helium [Ord+13].

A method to lower the freezing temperature of nitrogen is by mixing it with oxygen. The freezing point of this so called mixed refrigerant can be decreased to 50.1 K at 77 mol% oxygen and 23 mol% nitrogen. This could allow the economic operation of HTS applications in the 50 - 60 K region [KGO15].

Table 6.2: Boiling points and triple points of possible coolants for HTS cables

Cryogenic	Boiling Temperature at normal pressure / K	Triple Point Temperature / K
Helium	4.2	
Hydrogen	20.4	13.8
Neon	27.1	24.6
Nitrogen	77.4	63.2
Argon	87.3	83.8

6.2 Fault mitigation

For conventional cables the fault scenario is typically a short circuit for which design guidelines have been published in the *IEC 60364-5-54* standard. To the best of the authors knowledge no such guidelines have been published for superconducting cables. However an IEC standard for testing AC superconducting power cables is in the drafting phase.

*RE*BCO has a high intrinsic electrical resistivity in the normal conducting state. In consequence, the high electrical resistivity during an over current event would lead to a significant amount of joule heating within the cable. This heat influx could damage the superconductor (burn through, critical current degradation) or break down the dielectric.

One option to limit the heat influx is the use of a low resistive conductor in parallel connection, called stabilizer. As a conservative approach one assumes that the entire fault current flows in the stabilizer and not in the superconductor. The Joule heat is then

$$Q_{Joule} = R(T) \cdot I_{fault}^2 \tag{6.8}$$

with *R(T)* being the temperature-dependent electrical resistance of the stabilizer.

To calculate the resulting temperature rise during an over current event within the stabilizer an adiabatic energy balance is used

$$A \cdot L \int_{T_{Start}}^{T_{End}} c(T)\, dT = \int_{0}^{t_{fault}} \frac{\rho_{el}(T) \cdot L}{A} \cdot I_{fault}^2\, dt \tag{6.9}$$

With A being the stabilizer cross section, L cable length, $\rho_{el}(T)$ specific electrical resistivity of the stabilizer, $c(T)$ volumetric heat capacity of the stabilizer, and t_{fault} the fault duration. T_{start} is the starting temperature, and T_{end} the maximum temperature.

In [Muk+09] it was reported that a *RE*BCO power transmission cable was exposed to a fault current of $I_{\text{fault}} = 31500$ A for $t_{fault} = 2$ s. The majority of the current flowed over copper stabilizer with a total cross section of A_{stab} = 600 mm². During the fault the temperature increased from T_{start} = 77.3 K to T_{end} = 80 K.

Applying equation 6.9 to calculate the required copper cross section to reproduce the reported temperature rise

$$A_{stab} = \sqrt{\frac{\int_0^{t_{fault}} \rho_{el,stab}(T_{end}) \cdot I_{fault}^2 \, dt}{\int_{T_{op}}^{T_{End}} c_{stab}(T) \, dT}} \tag{6.10}$$

$$= \sqrt{\frac{\int_0^{t_{fault}} \rho_{el,stab}(T_{end}) \cdot I_{fault}^2 \, dt}{\int_{T_{start}}^{T_{End}} a_0 + a_1 \exp(a_3 T) \, dT}} \tag{6.11}$$

$$= 974\,\text{mm}^2 \tag{6.12}$$

with $a_0 = 3,513 \cdot 10^6 \,\frac{\text{J}}{\text{m}^3\text{K}}$, $a_1 = -5,414 \cdot 10^6 \,\frac{\text{J}}{\text{m}^3\text{K}}$ and $a_2 = -0,014 \,\frac{1}{\text{K}}$ and $\rho_{el,stab}(80\,\text{K}) = 2,276 \cdot 10^{-9} \,\frac{\Omega}{\text{m}}$.

According to Equation 6.10 974 mm² of copper would be required to achieved the observed temperature rise of the experiment performed in [Muk+09]. The actual stabilizer amount was 600 mm². The difference is likely caused by neglecting the heat capacity of all other cable components as well as as the heat transfer within the cable during the fault. Additionally the material properties are fitted in this case based on low temperature measurements of copper published in [SDR92; WC84] and therefore the properties might differ from the actual used materials.

6.3 Critical current calculation

The calculation of the critical current is complex due to its various dependencies (s. sec. 4.3). This results in varying critical currents over the conductor cross section (magnetic field, strain) and length (temperature, strain).

To simplify the problem it is assumed that the strain does not impact the critical current and that the temperature is constant. As a conservative approach the highest superconductor temperature within the system is chosen. With these assumptions the critical current calculation is simplified to solve the critical current magnetic field dependency of the superconductor. To simplify matters further, the problem is reduced to steady state and 2D which means that the critical current magnetic field dependency is expressed in the x-y-plane and that current flows perpendicular to the plane.

A fast implementation for calculating the critical current of a conductor is to approximate the conductor as a grid of infinitely long line conductors. In case a sufficiently dense grid is given, the magnetic field in a line conductor can be approximated by the sum of the magnetic fields produced by all neighboring line conductors. Applying the Biot-Savart law for infinitely long straight line conductors gives the magnetic field components of a line conductor $B_{i,x}$ and $B_{i,y}$ as

$$B_{i,x} = \frac{\mu_0}{2\pi} \sum_{j \neq i} I_c(B_{j,x}, B_{j,y}) \frac{(x_i - x_j)}{(x_i - x_j)^2 + (y_i - y_j)^2} \tag{6.13}$$

$$B_{i,y} = \frac{\mu_0}{2\pi} \sum_{j \neq i} I_c(B_{j,x}, B_{j,y}) \frac{-(y_i - y_j)}{(x_i - x_j)^2 + (y_i - y_j)^2} \tag{6.14}$$

with $I_c(B_{j,x},B_{j,y})$ being the critical current magnetic field dependence described in section 4.3.2 and equation 4.4. An implementation of this type was published in [ZQG16] and additionally provides an open source code written in MATLAB. Implementations of this type allow for rapid critical current calculations for e.g. stack like geometries. However complex geometries can quickly become challenging as the generation of line conductor grids can become very tedious.

It can therefore be beneficial to utilize the Finite Element Method (FEM). In commercial implementations of this method such as COMSOL, the geometries can be imported from computer aided design (CAD) software and the mesh generation is automated. Within such a software the magnetic vector potential

$$\nabla \times \left(\frac{1}{\mu} \nabla \times \boldsymbol{A} \right) = \boldsymbol{J} \tag{6.15}$$

In this case the J distribution is iteratively resolved with the constraint that $J = J_c(\boldsymbol{B})$ while the $\boldsymbol{B}$ field distribution across the tapes is refreshed in each iteration until self consistent $J_c(\boldsymbol{B})$ and $\boldsymbol{B}$ distributions are reached. For details on this method can be found in [Gri+14b].

Both implementations neglect the impact of varying contact resistances at the ends of the superconductor which could distort the current distribution. Therefore it is assumed that the current in each element of the calculation is at its critical point as defined by critical electrical field.

6.4 Electric insulation

The dielectric has the purpose of separating the surfaces at elevated voltage from its surroundings at ground potential. The ability of a material to prevent a discharge between two surfaces at different voltages is called dielectric breakdown strength. The breakdown of the dielectric is a statistical event. The break down voltages of a material is tested in a, preferably, large number of tests under a certain operational mode. The individual break down voltages are then fitted, usually with a Weibull distribution. Interpolation to a sufficient low breakdown probability (often 0.1 %) results in the breakdown strength of the material.

There are several factors that effect the breakdown strength of a material like electrode type, electrode surface roughness, dielectric volume etc. This unfortunately can lead to a rather wide spread of reported breakdown strengths.

According to [Bul67] the electric stress of a cylindrical DC cable can be calculated by

$$E(r) = \frac{U}{r \cdot \ln \frac{r_2}{r_1}} \tag{6.16}$$

with U being the rated voltage, r_1 and r_2 being the inner and outer diameter of the cylinder respectively. For the formula it is assumed that the temperature and the conductivity of the dielectric, is constant across the dielectric.

In superconducting cables two groups of dielectrics are often distinguished: Cold and warm dielectrics.

In cold dielectrics the insulation within the cryostat is at cryogenic temperatures. This has the advantage that the entire cryostat is at ground potential as well as allowing HTS screens which improves the performance of cable during transients or in AC operation. A typical cold dielectric used in cables is polypropylene laminated paper (PPLP) [GS12; Cho+12; Haz+02; Hwa+13; Kik+15; Kim+15; Kim+13a; Lee+13; Oku+02; Rez+10].

In a warm dielectric the insulation is applied to the outer wall of the cryostat at ambient temperatures. This allows the use of the full range of traditional dielectrics such as extruded cross linked polyethylene (XLPE). Furthermore, the inner cryostat can be, depending on the voltage, significantly smaller thus decreasing the cold surface and ultimately the heat leakage into the cryostat. The lower thermal mass within the cryostat can also be beneficial in terms of cool down time of the cable.

Limitations of the approach

At this point the accumulation of space charges in DC operation as well as various transient behaviors such as lightning impulses or polarity reversals are neglected. An overview of the electric fields of a DC cable in it's various operational stages can be found in [JM98].

The dielectric strength of the PPLP is slightly dependent on the pressure but saturates after 0.3 MPa [Kim+12]. There is also a slight difference between positive and negative polarity in which the positive polarity has a lower break down strength [Kim+12] and is therefore used as the reference polarity.

Insulating cooling channels

Section 6.1 describes the necessity of coolants to maintain the cryogenic temperatures. These coolants often flow in dedicated coolant channels that connect further to the cryogenic system. Generally, it is not desirable to operate the cryogenic system (valves, storage tanks, circulation pumps, etc.) at the operating voltage of the cable. In order to prevent this, the coolant can be surrounded by for example PPLP to electrically insulate the coolant channel from the remainder of the cable. Another approach could be to install electric insulation breaks [Kov+15] for example at the end of the cables.

6.5 Cable losses

Two types of losses are discussed within the frame of this work. First the losses due to heat leakage into the cryostat q_{cryo} and second electromagnetic losses within the superconductor q_{SC}.

$$q_{total} = q_{cryo} + q_{SC} \tag{6.17}$$

Cryostat losses

Cryostats are used to reduce heat leakage into the cold, superconducting system. Cryostats operating in the liquid nitrogen region (65 K - 77 K) are often multi walled vacuum vessels with multi-layer super insulation (MLI) in between. Multi layer insulation reduces the heat transfer by using multiple layer of reflective foil to reduce heat transfer by radiation. These layers are loosely stacked with spacers of low thermal conductivity thereby limiting the heat transfer by conduction. By placing MLI into a vacuum, the heat transfer by gas convection can be minimized.

Cryostats can be grouped into rigid cryostats and flexible cryostats. Flexible cryostats often utilize corrugated pipes to allow for bend ability and compensate for the thermal contraction when being cooled. Aside from the obvious advantage of a bendable cable, flexible cryostats also have the advantage that long cable pieces can be manufactured, evacuated and tested at the manufacturers facilities. The transport to the installation site can be achieved on drums. The flexibility of the cryostats comes at the price of higher heat transfer rates compared to rigid cryostats [Fes02b; Fes02a].

A challenge for both cryostat types is to guarantee a high vacuum for life times of 20 years and more. This is usually accomplished by installing getter materials into the cryostat that compensate for out-gassing of materials [Rey15, p. 200].

The calculation of heat transfer through multi layer insulation is very challenging due to unpredictable changes of parameters like winding pressure, uniform contact pressure or interstitial pressure [Neu04].

For this first conceptual design, experimentally established data is used. The thermal conductivity of several flexible, corrugated pipes with 30 to 60 layers of MLI was investigated in [Fes02a; Fes02b] at various pressures. It showed that a thermal conductivity of $k = 4$ mW/(m K) is achievable at pressures in the region of $p = 10^{-8}$ MPa for flexible cryostats that bridge room to liquid nitrogen temperature. With the thermal conductivity, Fourier's law can be used for a cylinder to calculate the specific heat load per meter q into the cold system

$$q_{cryo} = 2\,k\,\pi\,\frac{\Delta T}{\ln(d_{outer}/d_{inner})} \tag{6.18}$$

Applying this formula to a real cable for example the AmpaCity cable results in

$$q_{cryo} = 2 \cdot 0.4\,\frac{mW}{(m\,K)}\,\pi\,\frac{300\,K - 70\,K}{\ln(120\,mm/75\,mm)} = 1.2\,\frac{W}{m} \tag{6.19}$$

which is in the range of what has been reported for the cable [Sha+17].

Commercial flexible cryostats such as Nexan's Cryoflex [Nex] are usually only available at certain diameters which would impose additional restrictions on the design process. To avoid these restrictions it will be assumed that the outer diameter scales linearly with the inner diameter

$$d_{outer} = 1.5 \cdot d_{inner} + b \tag{6.20}$$

with $b = 13\,mm$. The formula is based on a linear regression of commercially available cryostats [Nex].

Electromagnetic losses

There are several sources of dissipation in a *RE*BCO tape. They were summarized in [Gri+14a] as follows

- Hysteresis losses, which are caused by the penetration and movement of the magnetic flux in the superconducting material
- Eddy-current losses, which are caused by the currents induced by a magnetic field and circulating in the normal metal parts of a superconducting tape
- Coupling losses, which are caused by the currents coupling two or more superconducting filaments via the normal conducting metal regions separating them
- Ferromagnetic losses, which are caused by the hysteresis cycles in magnetic materials

The hysteresis losses are the only mechanism that is unique to superconducting power applications and are often the dominant source for power dissipation. For this reason the other mechanisms are often neglected. The challenge in calculating the hysteresis losses is that they depend on the critical current of the tape. As already mentioned in section 4.3.2, the critical current of a tape not only varies with amplitude of the magnetic field but also with the angle. To accurately calculate the hysteresis losses, numerical models are required that have a sufficiently high resolution to consider these changes of the critical current.

There are several analytic models which help to provide a first estimate of losses within a *RE*BCO tape or cable under AC operation [NOR70; VM95]. For a tape or cable operating under DC current with a small ripple current no such analytic models have been developed as of yet. As a consequence some publication chose to adopt the pure AC models to the ripple currents under DC currents problematic as worst case scenario [Rey15, p. 210].

Investigations of the problem of AC ripple losses within a large DC current suggest that these losses are negligible compared to cryostat losses [Iva+16; Yos+14]. Under the assumption that high current application such aluminum electrolysis use very high pulse rectifiers [FI02], AC ripple losses will be neglected at this point.

In order to model the losses q_{SC} that a superconductor exhibits when it is operated close to critical current the power law will be used

$$q_{SC} = I_{op} E_c \left(\frac{I_{op}}{I_c} \right)^n \tag{6.21}$$

6.6 Strain

One of the main features of cables in comparison to bus bars is that they can be bent. Bending a cable leads to strain in most parts of the cable. The performance of the superconductor is strain-dependent therefore it is necessary to determine the bending limits that a cable can sustain.

One axis bending

For calculating the bending strain of a single HTS CroCo, the Euler-Bernoulli beam theory [Bal14] will be used, which is valid for axial dimensions far greater then radial (L >> R), small deformations, and one axis bending. The strain distribution is linear over the cross section with compression on the inner side and tensile strain on the outer side while a center line remains strain free (neutral axis). When comparing the strain limits of a *RE*BCO tape, one can see that the tensile strain limit is the weaker one and therefore limits the performance of the conductor [vE07]. With a linear strain distribution, the maximum strain acts at the edges of the CroCo.

$$\epsilon_{max} = \frac{D_{croco}}{D_{neutral}} \tag{6.22}$$

with D_{croco} being the diameter of the CroCo and $D_{neutral}$ the diameter of the bend neutral axis. For the investigated geometry it is important to know if the individual conductors can move or slip with respect to neighboring surfaces. When slipping is possible, the conductor bends around its own central axis. However, when it is fixed to a surface it bends around the center axis of that surface's body. As an example, the individual HTS tapes in a CroCo bend around the central axis of the CroCo because they are soldered together and therefore can not slip. This can also happen when for example a jacket exerts a high radial force on the CroCo.

For the case of slipping one can assume a maximum strain of $\epsilon_{max} = 0.4\%$ based on the strain limits shown in Table 4.4 and a CroCo diameter of $D_{CroCo} = 7$ mm (Sec. 5.2.4) are considered. This leads to a minimal neutral diameter of $D_{neutral} = 1.75$ m. This is sufficient to transport CroCos on conventional woods drums [Kab; Nex06].

For the no-slip case the minimum diameter of the drum can not be determined because D_{CroCo} is unknown in this case. However one can determine the maximum CroCo diameter $D_{\mathrm{CroCo,max}}$ that can be transported on the available drums which is 9.6 mm for a drum with 2.4 m inner diameter.

Helical bending

Section 5.2.1 describes the co-axial winding conductor in which *RE*BCO tapes are wound around a central core. The same principle can be applied to CroCos in order to modify the bending properties of cable consisting of CroCos or to mitigate thermal contraction during cool down (see Annex A.6).

When considering the CroCo for bending purposes it is assumed that it acts as a solid rod. This means when a CroCo is bend into a helical shape the outer tapes of the CroCo will always be under tensile strain, the inner tapes are under compression while the central tape is unaffected. The bending strain is then

$$\epsilon_{bending} = \frac{L_{max} - L_{neutral}}{L_{neutral}} \tag{6.23}$$

with

$$L_{neutral} = \sqrt{(\pi D_{neutral})^2 + p^2} \tag{6.24}$$

$$L_{max} = \sqrt{(\pi D_{max})^2 + p^2} \tag{6.25}$$

with D_{neutral} being the diameter to the central axis of the CroCo and p being the twist pitch of the helix and D_{max} is the maximum distance from the central axis of the CroCo to the edges of the cross.

Bending of helical wound cable

Bending a helical wound cable can result in additional strain acting upon the individual conductors. This can be avoided when the individual conductors can slip against each other [Li+13]. To ensure this behavior, the jacketing forces and therefore the friction forces need to be kept below a certain threshold. The determination of the friction forces is a whole process in itself. For a precise calculation of these please refer to literature [IEZ07; Pap97; HYL12; Li+13].

6.7 Design procedure

This section provides a step by step conceptual design process for *RE*BCO DC cables. A broad range of DC cables can be created based on the previously discussed formulas. However the high degree of freedom makes it difficult to define a practical design process without going to extraordinary lengths to cover all possibilities. Therefore additional constraints to the design process are imposed.

- The first constraint is the use of the Cross Conductor (CroCo) as superconducting strands in the cable. This is a design choice based on the observation that for the targeted very high currents the number of tapes increases into the hundreds. Combining the required superconductors into larger sub-strands allows for simplified cabling and offers the possibility of a modular design.
- The second constraint is the use of a dedicated channel for the coolant. This is a constraint to reduce the complexity of temperature and pressure calculations. Without a dedicated coolant channel the coolant would need to flow in the gaps between the CroCos which poses a problem of significant complexity in itself and is out of the scope of this work. For the purpose of this work, focus is put on a cable geometry with a central cooling channel.
- The third constraint is that the cable must be transportable on conventional drums.

The design process is illustrated in Figure 6.2. There are several possible ways to order the calculations steps while creating a viable processes. However, two of them appear to be logical, one can build a cable by starting with the inner most component and build outwards or one can start at the outer limit and build inwards. The first approach would likely be faster in finding compact designs while the second would likely be faster in finding cost efficient designs.

The process displayed in Figure 6.2 starts from the outer most layer and builds inwards.

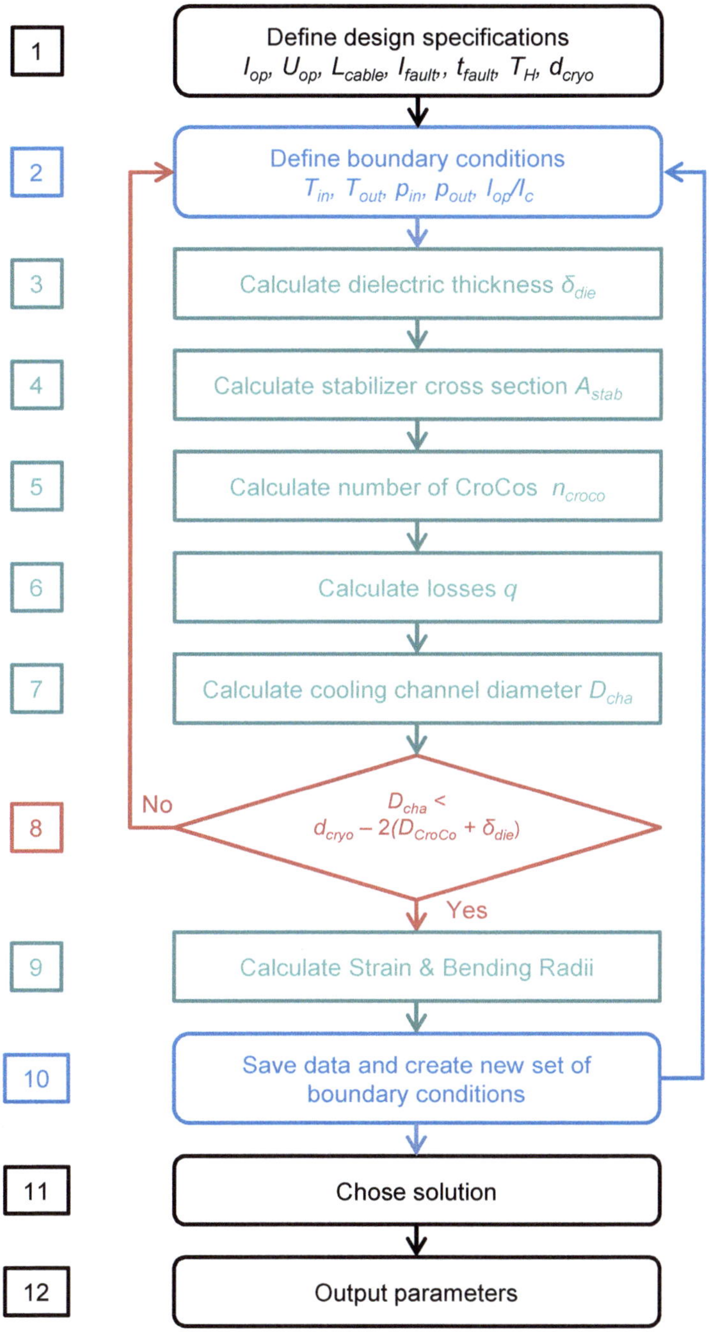

Figure 6.2: Design process for conceptual *RE*BCO DC cable utilizing HTS CroCos with liquid nitrogen cooling

Step 1: Design specifications

The design specifications define the required operational parameters for the target application and have to be defined as a first step in the design process. Table 6.3 lists the minimum design specifications.

Table 6.3: Required design specification for *RE*BCO DC cable

Description	Symbol
Operational current	I_{op}
Operational voltage	U_{op}
Cable length	L
Fault current	I_{fault}
Fault duration	t_{fault}
Ambient temperature	T_{H}
Inner cryostat diameter	d_{cryo}

In addition to these design specifications material properties and parameters need to defined. This includes coolant and dielectric properties.

The design specification can also be extended by boundary conditions to reduce the parameter space in an optimization process. As an example, the coolant could be defined as a design specification or certain geometrical structure could be imposed.

The inner cryostat diameter is of particular importance, as it acts as limiting geometric condition which often decides if a set of specifications has a solution.

Step 2: Boundary conditions

In addition to the design specifications, certain boundary conditions need to be defined in order to solve the equation system and generate a conceptual design. In contrast to the design specification, the boundary condition can be modified during an optimization process. Table 6.4 lists the boundary conditions that were chosen for this design process. The boundary conditions were chosen due to the fact that it allows more freedom within the design process if the temperatures and pressures are defined as boundary conditions. This way it is possible to perform for example critical current calculations before the thermal design is completed.After this step, the cable consist of a cryostat and will be filled in the next steps. This is schematically shown in Figure 6.3. The size of the outer cryostat is only a suggestion at this point and may vary.

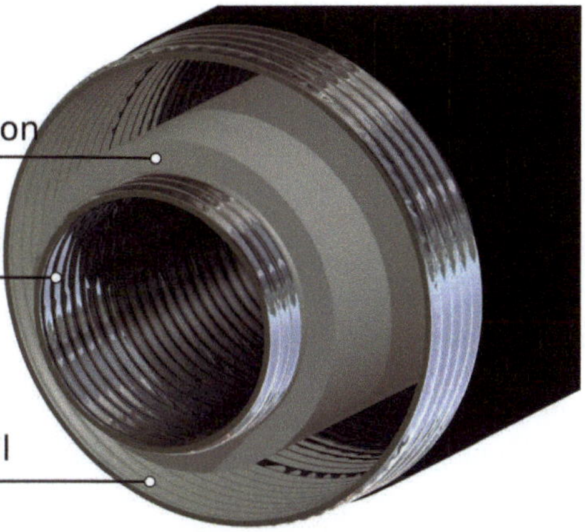

Figure 6.3: Schematic display of a conceptual cable design after Step 2 of the design process

Table 6.4: Required boundary conditions

Description	Symbol
Inlet temperature	T_{in}
Outlet temperature	T_{out}
Inlet pressure	p_{in}
Outlet pressure	p_{out}
Operational to critical current ratio	$I_{\text{op}}/I_{\text{c}}$

Step 3: Dielectric layer

In this step the dielectric thickness is calculated to insulate the superconducting cable core from the inner cryostat. Solving Eq. 6.16 at $r = r_{die}$ with $r_2 = r_{die} + \delta_{die}$ gives the thickness of the dielectric layer δ_{die}

$$\delta_{die} = r_{die}\left(\exp\left(\frac{S \cdot U}{E_{BD} \cdot r_{die}}\right) - 1\right) \tag{6.26}$$

To ensure that the electric field is greater than the breakdown strength, the safety factor S is always larger than one. In high current operation the voltages are often relatively low (< 2000 V), this can result in a very thin dielectric. In case the dielectric is realized by several tapes wound around the core it needs to be considered that several layers are required to achieve a sufficient overlap between the tapes. For PPLP the minimum layer count is usually between 3 and 5 layers.

In case of a warm dielectric cable the dielectric will be applied to the outer wall of the warm cryostat wall.

Step 4: Stabilizer cross section

All variables in the stabilizer calculation are defined in the design specifications and boundary conditions. For small temperature differences the stabilizer cross section can be quickly calculated with Eq. 6.10. A possible procedure for large temperature differences is discussed as an example in Section 6.8.

Step 5: HTS CroCos

If the inner diameter of the dielectric layer is known, the maximum number of CroCos that can fit into one layer following the dielectric can be calculated by dividing the available circumference by the CroCo diameter

$$N_{max,CroCo} = \frac{\pi \left(d_{dielectric} - D_{CroCo}\right)}{D_{CroCo}} \tag{6.27}$$

with D_{CroCo} being the outer diameter of the CroCo. Dividing the required critical current by the number of CroCos provides the required critical current per CroCo

$$I_{c,CroCo} = \frac{I_{c,cable}}{N_{max,CroCo}} \tag{6.28}$$

The number, location and critical current of the cable are known and the operational temperature is defined in Step 2. Therefore critical current calculations by the methods outlined in section 6.3 are possible. This is an iterative process in which the critical current of the CroCos is adjusted until the cable critical current is met. This can be done for example by changing the number of *RE*BCO tapes per CroCo or changing the tapes properties (magnetic field behavior, critical current). The critical current values listed in Table 5.4 can act as initial values in this process.

During the iterative CroCo design process one should consider the required amount of stabilizer material calculated in Step 4. It can be beneficial when all the necessary stabilizer is already build into the CroCo. Otherwise, the lacking stabilizer material needs to be added for example into the voids between adjacent CroCos and connected to the electrical circuit.

Should it not be possible to achieve the required critical cable current with a single layer of CroCos than additional layers can be added.

After this step the conceptual design features the cryostat, dielectric and the superconducting part in form of the CroCos. A schematic of the status at this point is shown in Figure 6.4. The final component of this design process will be a cooling channel in the center of the cable.

Figure 6.4: Schematic of a possible conceptual cable design after calculating the number of CroCos

Step 6: Loss calculation

The cable losses can be calculated based on the design specifications and the boundary conditions. The calculated losses include the electromagnetic losses of the superconductor modeled by the power law as well as heat leak through the cryostat

$$q = 2\,k\,\pi \frac{\Delta T}{\ln\left(1,5 + \frac{b}{d_{cryo}}\right)} + I_{op} E_c \left(\frac{I_{op}}{I_c}\right)^n \tag{6.29}$$

with $k = 4$ mW/(m K) being the experimentally determined heat transfer through the cryostat and $b = 13\ mm$ a fit factor (section 6.2 and 6.5).

Step 7: Cooling channel dimensions

Based on the boundary condition of a central round cooling channel, the hydraulic diameter is set to the diameter of the cooling channel $D_{cha} = D_h$. For the purpose of this conceptual design a corrugated pipe will be used as cooling channel.

The determination of the Darcy friction factor for corrugated pipes is non-trivial and outside of this work. Therefore a worst case assumption is made based on a meta study on friction factors of corrugated pipes [Ji+15] and set it to be 4.5 times the friction factor of smooth pipes. This allows the use of well developed empirical equations for smooth pipe friction factors [RRM02]. These equations are dependent on the Reynolds numbers and therefore on the fluid velocity resulting in an implicit problem.

The temperature rise due to the pipe friction can be rewritten with pressure drop resulting in

$$\Delta T = \frac{\Delta p}{\rho \cdot c_p} + \frac{q \cdot L}{A_{cha} \cdot v \cdot \rho \cdot c_p} \tag{6.30}$$

with and $\dot{m} = A_{cha} \cdot v \cdot \rho$.

In order to eliminate the coolant velocity Eq. 6.6 is inserted as

$$v = \sqrt{\frac{\Delta p \cdot D \cdot 2}{\zeta \cdot \rho \cdot L}} \tag{6.31}$$

and solved for A_{cha}

$$A_{cha} = \frac{q \cdot L}{(\Delta T \cdot \rho \cdot c_p - \Delta p) \cdot v} \tag{6.32}$$

$$\frac{\pi D_{cha}^2}{4} = \frac{q \cdot L \cdot \sqrt{\zeta \cdot \rho \cdot L}}{(\Delta T \cdot \rho \cdot c_p - \Delta p) \cdot \sqrt{\Delta p \cdot D_{cha} \cdot 2}} \tag{6.33}$$

$$D_{cha} = \left(\frac{4 \cdot q \cdot L \cdot \sqrt{\zeta \cdot \rho \cdot L}}{\pi \left(\Delta T \cdot \rho \cdot c_p - \Delta p\right) \cdot \sqrt{\Delta p \cdot 2}} \right)^{2/5} \tag{6.34}$$

Eq. 6.34 calculates the required inner channel diameter for the coolant channel to satisfy a given temperature and pressure drop along the cable. Due to the Darcy friction factor ζ this equation needs to be solved numerically.

Step 8: Geometric compatibility check

In the previous step the diameter of the coolant channel was calculated. The available space within the cable is restricted by dielectric, the stabilizer material and the CroCos as calculated in Step 5. In this step the remaining space is compared with the available space. If the available space is not sufficient than the current set of boundary conditions does not have a solution.

$$D_{cha} < d_{cryo} - 2 \cdot (D_{CroCo} + \delta_{die}) \tag{6.35}$$

Depending on the magnitude of discrepancy one might reconsider the CroCo design.

After a fitting coolant channel is found all components of the conceptual design are created. A possible conceptual design at this point is shown in Figure 6.5

Step 9: Strain and bending Radii

To transport the cable on a drum the CroCos need to be helical wound around the core. The minimum twist pitch of this helix is calculated by

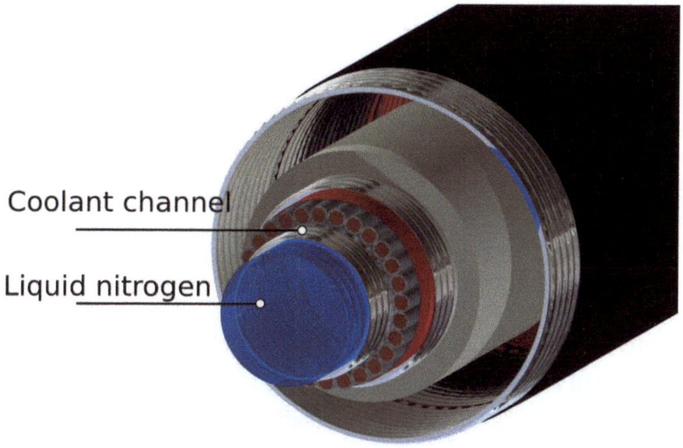

Figure 6.5: Schematic of a conceptual design with all components present

$$p_{min} = \frac{\pi\sqrt{D_{max}^2 - D_{neutral}^2(1 - 2\epsilon - \epsilon^2)}}{\sqrt{\epsilon^2 + 2\epsilon}} \tag{6.36}$$

with ϵ being the degradation strain limit of the used superconductor.

Step 10: Save data and create new set of boundary conditions

Based on the optimization goal, a new set of boundary conditions can be chosen for which a new design will be generated.

Step 11: Chose solution

A solution is chosen based on the optimization goal out of all possible solutions.

Step 12: Out parameters

All calculated parameters are returned when the final design is chosen.

6.8 Design study of 35 kA *RE*BCO DC cable demonstrator

A fully developed cable prototype including all components described in the previous section would significantly exceed project resources. In this project the focus is on the superconducting component of the cable. The design process is therefore simplified as the cooling is reduced to an open liquid nitrogen bath.

Step 1: Specifications

The specifications for the cable demonstrator are listed in Table 6.5. The operational current of 35 kA was determined as a first step towards a 100 kA-class cable.

The operational voltage was limited due to safety concerns to less than 50 V. The cable length was defined to be at least 3 m.

The fault current is defined by the maximum current of the power supply which is 50 kA. The dedicated quench protection used in the project is expected to detect a quench and shut down the power supply within 1 s. More information on current source and quench detection can be found in 8.2.3.

The operational temperature is set to the saturation boiling temperature at normal pressures of liquid nitrogen.

The maximum outer diameter was set to 130 mm. This is derived from the fact that larger diameters significantly reduce the amount of cable that fit on standard drums. Therefore it appeared sensible to limit the diameter to dimensions that are relevant for a full scale cable.

Table 6.5: Summary of cable demonstrator specifications

Description	Symbol	Value
Operational current	I_{op}	35000 A
Operational voltage	U_{op}	<50 V
Cable length	L	3 m
Fault current	I_{fault}	50000 A
Fault duration	t_{fault}	1 s
Operational temperature	T_{op}	77 K
Maximum outer diameter	D_{max}	130 mm

Step 2: Boundary conditions

Due to the simplified design process the only boundary condition necessary to solve the system of equations is the maximum permissible temperature rise during fault T_{end}. In order to prevent degradation of superconductors during faults, the maximum temperature should not exceed 400 K

[Kim+08]. For this project the maximum temperature during fault was set to be T_{end} = *323 K*. This is possible as there is no dielectric that could be compromised by gaseous nitrogen. Therefore the main risks are damage to the superconductor, the cable or measurement setup. The measurement setup and the cable could be damaged if for example soldering connections dissolve. To prevent this, only solder with melting temperatures above 323 K should be used in this project (see Table 7.1).

Step 3: Dielectric

No dedicated dielectric was required due to low voltage requirement. In order to minimize leak currents the test setup was insulated as described in section 8.2.

Step 4: Stabilizer cross section

Copper is chosen as stabilizer material due to its good solderability, electrical resistance, heat capacity as well as price. The electrical resistance and the heat capacity of copper as a function of the temperature are summarized in Annex A.2. The material properties and the boundary conditions are sufficient to solve the adiabatic energy balance (Eq. 6.10) under the assumption of $\rho_{el,stab}(T_{end})$. For a 50 kA fault current, a fault duration of 1 s and an end temperature T_{end} = 323 K, the required stabilizer cross section is A_{stab}= *252 mm²*.

The difference between operating temperature and end temperature is with $\Delta T = 296\,\text{K}$ large. Consequently the difference between the electrical resistance at the start of the fault $\rho_{el,stab}(77\,\text{K}) = 2 \cdot 10^{-9}\,\Omega\text{m}$ is about one order of magnitude smaller compared to the resistivity at the end $\rho_{el,stab}(373\,\text{K}) = 2,2 \cdot 10^{-8}\,\Omega\text{m}$. Therefore the stabilizer cross section calculated by Eq. 6.10 appears to be significantly overestimated.

One way to mitigate this problem is to solve equation 6.9 in small increments of the temperature gradient for the time it takes to achieve the temperature increment. This results in a series of time increments which in sum is the total fault duration. Now the stabilizer cross section can be altered until the total fault duration matches the requirements.

The advantage of this process is that the electrical resistance is only constant over a small temperature increment. Thus resulting in a more accurate estimation of the required stabilizer cross section for large temperature differences. Figure 6.6 shows the temperature of the stabilizer as a function of time at a fault current of 50 kA calculated with the previous described process. At a stabilizer cross section of A_{stab} = *169 mm²* it takes longer than 1 s to reach the end temperature and therefore a copper cross section of 169 mm² is sufficient in this fault scenario.

An implementation of above described process written in MATLAB is attached in Annex A.3.

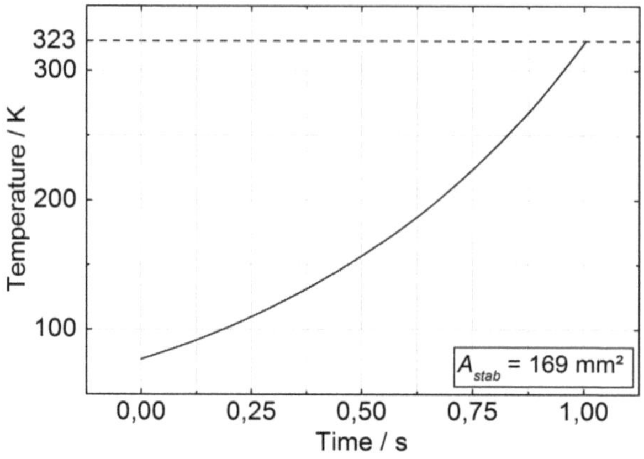

Figure 6.6: Temperature development within the stabilizer material as a function of time at a fault current of 50 kA with a stabilizer cross section of 169 mm².

Step 5: HTS CroCo

Section 5.2.4 describes three CroCo variations with different outer diameters. For the purpose of this design some space for cabling and jacketing is added on top of the outer diameters resulting in outer diameters of 10 mm, 8 mm and 6 mm. Calculating the maximum number of CroCos that fit into the 130 mm cable results in: 37, 47 and 64 CroCos for the different CroCo diameter respectively. For the critical current calculation the number of CroCos is reduced to 36, 45 and 60 respectively. This is a design choice based on the understanding that manufacturing and assembly of the cable demonstrator is easier when the number of CroCo is a integer fraction of 360°.

The operational current of the demonstrator cable is in this design process equal to the critical current.

The specification of the 35-kA cable demonstrator dictated that the demonstrator should be easily upgradable towards a 100-kA class cable. Therefore the critical current calculations were performed for a 100-kA cable and than down scaled based on feasible design for a 35-kA class cable. The critical current calculations were performed in accordance to the methods described in section 6.3 and performed with the commercial FEM software COMSOL.

Through an iterative process, one CroCo design per size was determined which would achieve 100 kA at 77 K s.f. under the previously described limitations. The results of the iterative process are summarized in Table 6.6. The first section of the table describes the structure of the CroCos. The *RE*BCO tapes were assumed to be 85 μm thick and their electroplated copper was neglected in the calculation of the amount of copper per CroCo which only refers to copper added on top of the electroplated copper. As can be seen the largest CroCo still contains a significant amount of copper while the smallest one does not contain any additional copper. This is due to the smaller cross section available for the smaller CroCos which is caused by the condition of a single layer of CroCos. Thus enforcing more tightly packed *RE*BCO tapes in the smaller CroCo types.

Table 6.6: Summary of iterative CroCo design to achieve 100-kA class cable consisting of CroCos of various sizes

		CroCo properties		
CroCo type		6/4	4/3	3/2
*RE*BCO tape number		22 x 6 mm 10 x 4 mm	26 x 4 mm 8 x 3 mm	24 x 3 mm 12 x 2 mm
Copper cross section per CroCo		17.2 mm²	4.62 mm²	0
Diameter		7.3 mm	5.2 mm	3.6 mm
Distance between *RE*BCO layers		185 μm	120 μm	85 μm
		Cable properties		
Number of CroCo		36	45	60
Tape I_c sum (I_{c0}) 77 K s.f.		186 kA	173 kA	173 kA
	SST	104 kA	97 kA	96 kA
I_c, 77 K s.f.	DNA	110 kA	103 kA	102 kA
	Theva	111 kA	104 kA	103 kA
Number of CroCo		12	15	20
Tape I_c sum (I_{c0}) 77 K s.f.		62 kA	58 kA	58 kA
	SST	37 kA	34 kA	34 kA
I_c, 77 K s.f.	DNA	40 kA	37 kA	37 kA
	Theva	40 kA	37 kA	36 kA
Total copper cross section		206 mm²	69 mm²	0

The second part of the table shows the critical current of each CroCo size in a 100 kA arrangement. The critical currents were calculated with three parameter sets of the critical current magnetic field dependence that are typical for the listed manufacturers (s. Figure 4.7). A critical current of 30 A/mm_{width} was assumed for all *RE*BCO tapes and calculations. The sum of the critical current of all *RE*BCO tapes under self field conditions is listed ($I_{c,0}$). The I_c/I_{c0} ratio is a measure for how efficient the *RE*BCO tape is used and it is nearly constant across the different CroCo types but varies between about 0.56 to 0.6 for the tapes from different manufacturers.

The third part of table shows the critical current when the number of CroCo is reduced to one third of the 100 kA scenario. This is the target scenario of this design process. The I_c/I_{c0} remains nearly constant across CroCo types but increases from 0.59 to 0.64 across different manufacturers. This increase is due to the decrease in the magnetic field as can be seen in Figure 6.7.

Figure 6.7 shows the magnetic field distribution of the 6/4 type CroCo in the two scenarios investigated. In Figure 6.7 only one CroCo is shown as the remaining CroCos were replaced by symmetry conditions within the calculation in order to save computational time. The left figure shows a tightly packed 100 kA-class arrangement of 36 CroCos. The right figure shows a down scaled arrangement which would achieve 37 kA and consists of 12 CroCos.

Figure 6.8 is a detail of Figure 6.7 and show the magnetic field distribution within a CroCo. The arrows indicate the direction of the magnetic field. It can be seen in the 104 kA arrangement that the magnetic field increases towards the upper half of the CroCo and peaks at $B = 0.44\ T$ in the outer *RE*BCO tapes. While the magnetic field in the bottom half is significantly reduced. In the down scaled arrangement the low field region is almost in the center of the CroCo, suggesting very little impact from neighboring CroCo. Also the peak field in the outer layer is significantly lower with $B = 0.28\ T$. This also shows in the critical currents. The CroCo within the down scaled arrangement has an individual critical current of $I_{c,CroCo,cable}$ = 3148 A (SST) while it achieves $I_{\mathrm{c,CroCo,s.f.}}$ = 3114 A (SST) under self field conditions.

The magnetic field distribution of smaller CroCo types are nearly identical with peak magnetic fields of $B = 0.42\ T$ and $B = 0.43\ T$ for the 4/3 and 3/2 type respectively in the 100 kA arrangement. In the down scaled arrangement the peak fields are $B = 0.29\ T$ (4/3 type) and $B = 0.30\ T$ (3/2 type) are higher than in the 6/4 type. The cause for the higher field in that scenario is the tighter arrangement of the *RE*BCO tapes within the CroCo.

Step 6: Losses

The losses are being discussed as part of the description of the test setup in Section 8.2.1.

Step 7 & 8: Coolant channel & geometric compatibility check

The cable will be tested in an open liquid nitrogen where no temperature or pressure drop is expected. Therefore these steps are not necessary.

Step 9: Strain and bending radii

The strain and bending properties of the demonstrator cable will investigated in future work.

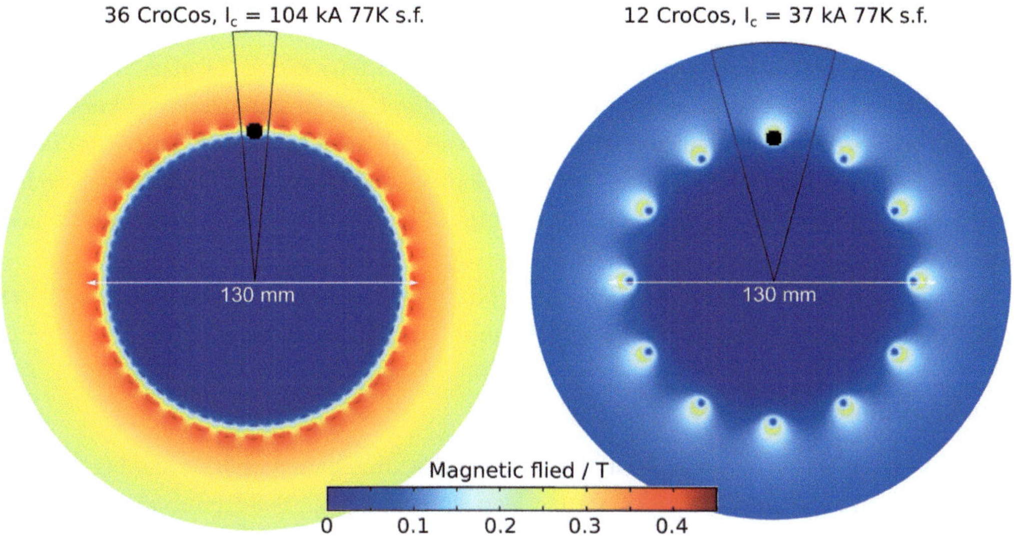

Figure 6.7: Magnetic field distribution of 6/4 CroCo in a 104 kA arrangement and a down scaled 37 kA arrangement.

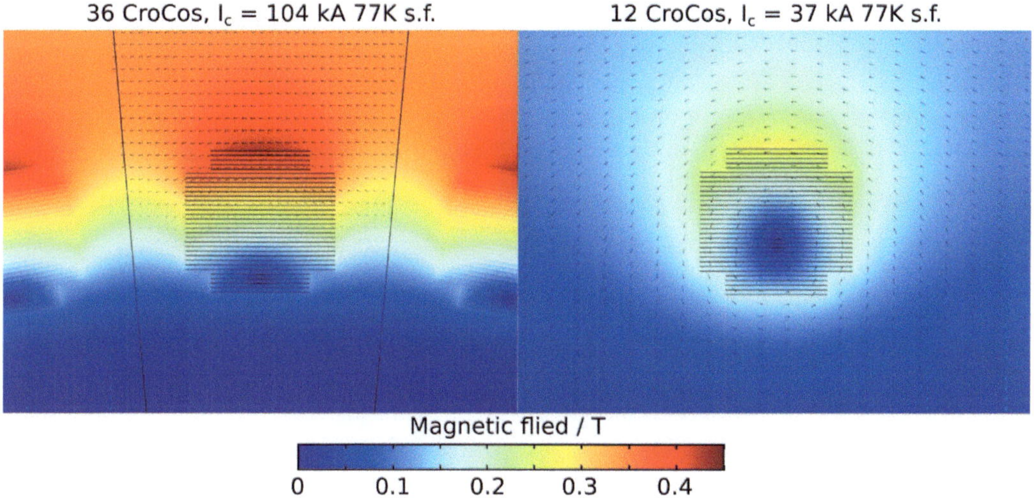

Figure 6.8: Detailed magnetic field distribution of 6/4 type CroCo104 kA arrangement and a down scaled 35 kA arrangement

Step 10: Save data and create a new set of boundary conditions

Since the CroCo contains sufficient amount of stabilizer no further iteration is required.

Step 11: Chose solution

It is possible to create a single layer 100-kA-class cable with CroCos of all sizes. However in case of the 4/3 and 3/2 CroCo arrangement significant amounts of stabilizer would need to be added in order to meet requirements. The 6/4 CroCo arrangement has already build in sufficient amounts of stabilizer to withstand the specified short circuit. The larger CroCos are also less costly due to increased losses during the tape cutting which makes smaller tapes usually more expensive.

In conclusion the 6/4 arrangement appears to be the logical choice and was therefore chosen in this design process. The 35 kA cable will therefore consist of 12 6/4 CroCos.

Step 12: Output parameters

To achieve a critical current of at least 35 kA, 12 CroCos are required each containing 22 6 mm wide tapes and 10 4 mm wide tapes with at least 30 A/mm_{width}. The tape thickness was set to 185 µm of which 100 µm are designated as copper stabilizer. The CroCos will need to be mounted onto a cable core with a diameter of 110 mm and operated in liquid nitrogen in order to reach the expected critical current.

7 HTS CroCo Manufacturing

7.1 Thermal stability of *RE*BCO tapes

The soldering process is the key process of the HTS CroCo manufacturing process. Within the soldering process *RE*BCO tapes are exposed to elevated temperature for a certain amount of time. In [Yaz+13] the critical current degradation of two different *RE*BCO tapes was investigated for tapes that were exposed to 135°C and 195°C for up to 36 h. No degradation was found at 135°C. At 195°C significant degradation was found in both tapes at different duration.

This degradation limits the number of available solders when additional properties like flow characteristics during soldering, mechanical or electrical properties at cryogenic temperatures need to be considered. The solder selection becomes increasingly challenging when manufacturing requires multiple soldering steps. This often necessitates several solders with staggered melting points to prevent the melting of previous solder steps. Table 7.1 lists basic properties such as electrical resistance and melting point for solders that have been used with *RE*BCO tapes. The price of the solders varies depending on the global market situation of expensive metals such as silver or indium but it illustrates the orders of which they differ. To estimate the cost of solder in a CroCo, a 6/4 CroCo with a solder matrix diameter of 8 mm and a filling factor of 75 % is considered. When using for example In-Sn, the solder cost would amount to about 75 €/m while Sn-Pb would only cost about 2 €/m.

Lead containing solders belong to the hazardous substance according to European legislation (Restriction of Hazardous Substances, 2011/65/EU) making it challenging to use them outside the scientific community.

Better knowledge of the degradation behavior could enable the use of alternatives such as Sn-Ag for some soldering processes. The following results were published in [Pre+18].

Table 7.1: Properties of solder used for *RE*BCO tapes[Lu+11; Ted+12; Zha+11; BBW14]

Solder	Composition / m%	T_l / °C	ρ(77 K) /$\mu\Omega$cm	ρ(293 K) /$\mu\Omega$cm	Price / €/kg
In-Bi	In66Bi33	72	15	-	1000
In-Sn	In52Sn48	118	9	-	800
Bi-Sn	Bi57Sn43	139	-	38	80
In	In97Ag3	141	-	-	1250
Sn-Pb	Sn63Pb37	183	2.5	14	20
Sn-Ag	Sn96Ag4	221	2.5	12	50

7.1.1 Experimental setup and procedure

To reproduce soldering conditions, an aluminum solder bath was used (s. Fig. 7.1). The dimensions of the liquid solder volume of the bath are 10 cm x 1.5 cm x 1.5 cm. The commonly available Sn63Pb37 was used as solder. The heat was provided by two heat cartridges. The temperature of the solder was independently checked by a Type K thermocouple immersed into the liquid solder.

The critical current measurements were performed with a Keysight 6671A DC source, a Keithley 2000 Multimeter coupled with a Femto DLPVA preamplifier at 60 dB power amplification.

Spring-loaded pins with 7 cm distance in between were used as voltage taps. The taps were positioned on the HTS side of the *RE*BCO tapes. All critical current measurements were performed in an open LN_2 bath.

The critical temperature T_C was determined by four-probe resistance measurements using a physical property measurement system (PPMS, from Quantum Design). The electrical resistance was determined through voltage measurement with a current of 1 mA.

All *RE*BCO tapes were cut from spools to 15 cm long samples and cleaned with Isopropanol. Subsequently, their initial critical current was determined based on an electric field criterion of 1μV/cm.

For the heat treatments, the samples were placed into the solder bath for a certain amount of time. Approximately 2 cm on each end of the samples were outside the solder bath. No flux was used.

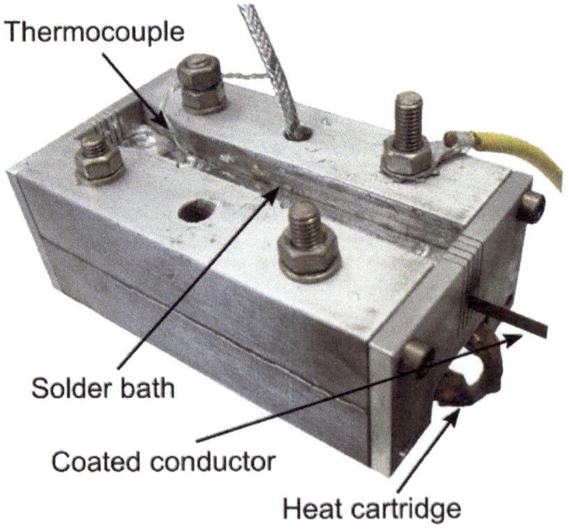

Figure 7.1: Soldering bath with a *RE*BCO tape inserted

The solder bath temperature was kept constant within ±2.5°C. Up to three samples were heat-treated in the solder bath simultaneously. After the samples were removed from the solder bath, they were cooled to room temperature and subsequently measured again in LN_2.

*RE*BCO tapes from SuperPower Inc. (SPI), Shanghai Superconductor Technology (SST) and Deutsche Nanoschicht (DNA) were investigated. In Table 7.2, the sample properties are summarized. All samples use electroplated copper of various thicknesses as stabilizer material. Unless otherwise noted, the SPI samples were from batch M4-154-2 FS. To allow comparison, the results for each tape were normalized to their initial, untreated critical current $I_c/I_{c.0}$.

Table 7.2: Summary of sample properties from Superpower Inc. (SPI), Shanghai Superconductor Technology (SST), and Deutsche nanoschicht (DNA)

Company	Batch ID	Width / mm	Stabilizer / µm	Year of delivery	$I_{c,min}$ / A
SPI	M4-69-4 BS	4	100	2013	107
SPI	M4-154-2 FS	4	40	2014	85
SPI	M3-1072-3 0306	4	110	2014	115
SPI	M3-1050-4 0103	3	40	2016	76
SST		4	40	2016	120
DNA		10	40	2017	250

7.1.2 Results

Accumulation of critical current degradation

In a first step, it was investigated if the number of heat treatments has an impact on the critical current degradation. In Figure 7.2 the normalized critical currents of several samples that underwent a series of subsequent heat treatments of various duration's is shown. The x-axis shows the accumulated time the samples spent in the solder bath. The solder bath temperatures were 225 °C and 250 °C for Fig. 7.2 (a) and (b), respectively.

In Fig. 7.2 (a) it can be seen at 5 min that the relative critical current of the four samples is almost the same, even though samples 26 and 27 where brought to 5 min total by a series of three 1 minute and one 2 min heat treatment. On the other hand sample 28 and 29 had just one 5 min heat treatment. Similar behavior can be observed at 15 min.

In Fig. 7.2 (b) this independence from the number of heat treatments can also be observed for a 250 °C heat treatment.

Subsequently it was assumed that this behavior is also valid for all other temperature levels investigated (200 °C, 275 °C, 300 °C). In consequence, samples were used multiple times to reduce

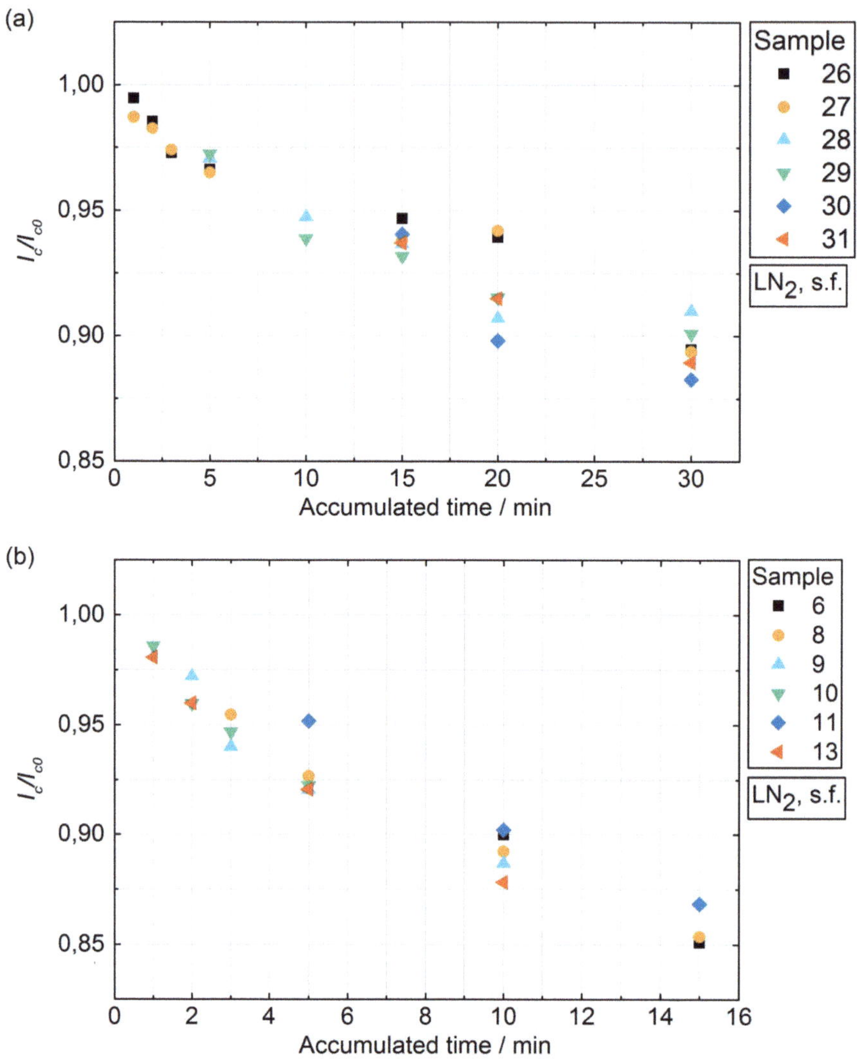

Figure 7.2: Critical current degradation over multiple heat treatments at (a) 225 °C and (b) 250 °C for the SPI tape.

the amount of samples necessary for this type of investigations. In addition all investigations used the accumulated time of the samples.

Different production batches

In order to determine if the results are representative it needs to be investigated if the critical current degradation under soldering conditions depends on the production batch. This is also important for larger applications like cables that require multiple production batches of *RE*BCO tapes. The degradation of four different production batches from SPI tapes were investigated, two at 225 °C and two at 250 °C. The results are shown in Fig. 7.3.

No difference between the production batches was observed at the respective temperatures. The exact date of manufacturing is not known however the orders are from 2013 to 2016 indicating that this behavior has been fairly constant for the last years.

The investigated tapes in Fig. 7.3 have a slightly different structure. The samples measured at 250 °C have an overall stabilizer thickness of 110 µm and 100 µm, respectively. The tapes investigated at 225 °C have a stabilizer thickness of 40 µm (s. Table 7.2). To investigate the influence of different stabilizer thickness, one batch with 40 µm thick stabilizers is compared with tapes having 110 µm and 100 µm thick stabilizers in Figure 7.4. For duration <10 min the degradation of all three batches appears to be similar. For duration >20 min the degradation of the tapes with 40 µm stabilizer is noticeably less than the tapes with 100/110 µm stabilizer batches.

Due to unknown details of the manufacturing process a cause for this behavior could not be found.

Critical current degradation as a function of temperature and time

The critical current degradation of a *RE*BCO tapes under soldering conditions shows a pronounced dependence on time and temperature as shown in Fig. 7.5. At 250 °C a degradation of 10% occurs after 10 min. At higher soldering temperatures of 300 °C it is visible that the same degradation of 10% occurs already after 1 min. After 10 min at 300 °C the critical current, reaches only 50 % of the initial current. Figure 7.6 displays the critical current degradation as a function of time up to 120 min and for various temperature levels between 225 °C and 300 °C.

The strong temperature dependence can be illustrated by the 95 % I_c retention criteria. At 225 °C it takes about 10 min for the tape to degrade to 95 % of its initial critical current. At 250 °C, 275 °C and 300 °C it only takes about 5 min, 1 min and 30 s, respectively.

Comparison between different manufacturers tapes

In addition to tapes from SuperPower, samples from Shanghai Superconductors and Deutsche Nanoschicht were investigated for duration up to 10 min which turned out as a critical time for the 95 % retention criteria. In Figure 7.7 it can be seen that the degradation of SPI and STT is very similar. The samples of DNA showed a stronger degradation for soldering times larger than 2 min at 250 °C compared to the other manufacturers.

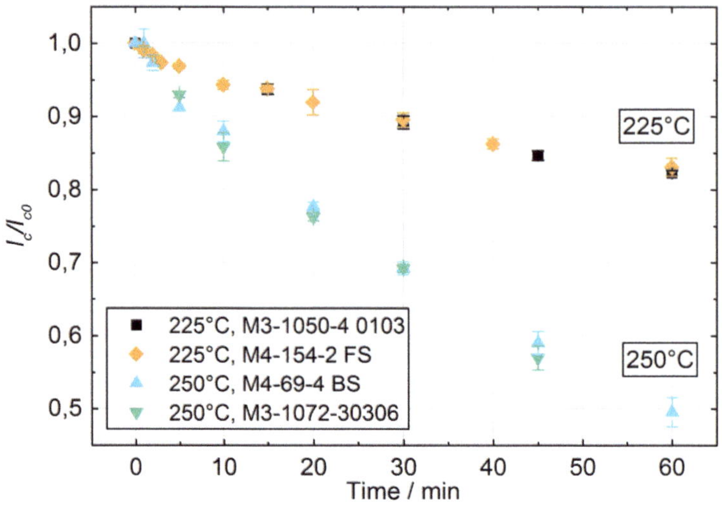

Figure 7.3: Critical current degradation with varying production batches for SPI tapes

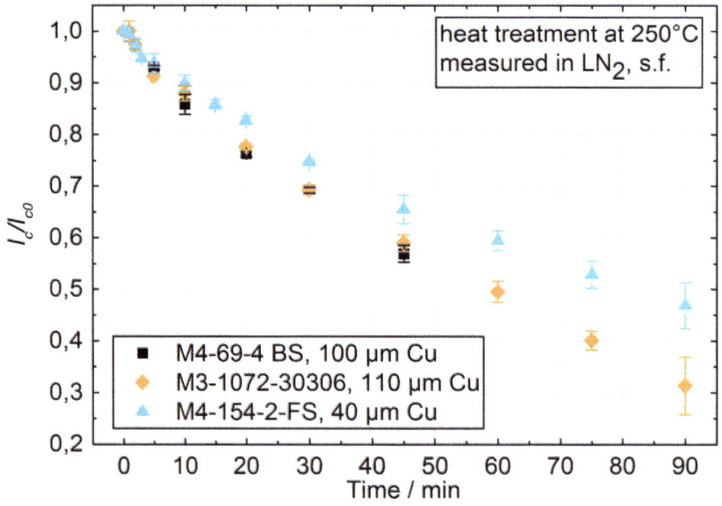

Figure 7.4: Critical current degradation of multiple production batches at 250°C with different copper stabilizer thicknesses of SPI-tapes.

Critical temperature measurements

A likely cause for the degradation is oxygen diffusion out of the crystal structure of the *RE*BCO. Localized measurements of the oxygen content in *RE*BCO tapes are difficult. Therefore, the

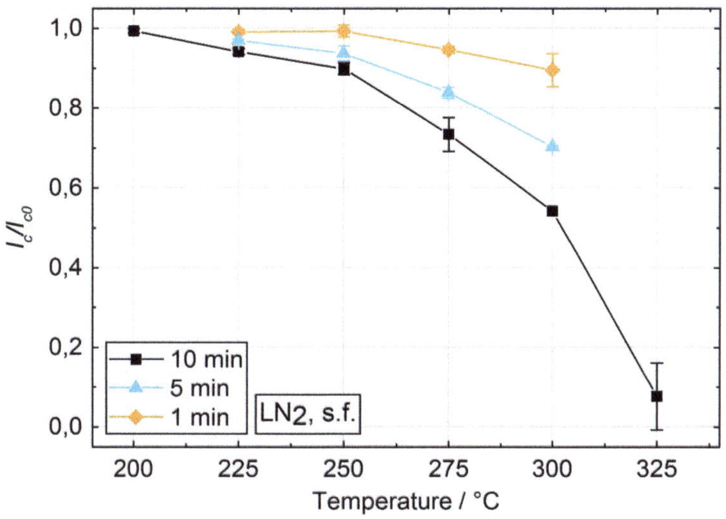

Figure 7.5: Critical current degradation of SPI tapes as a function of the temperature.

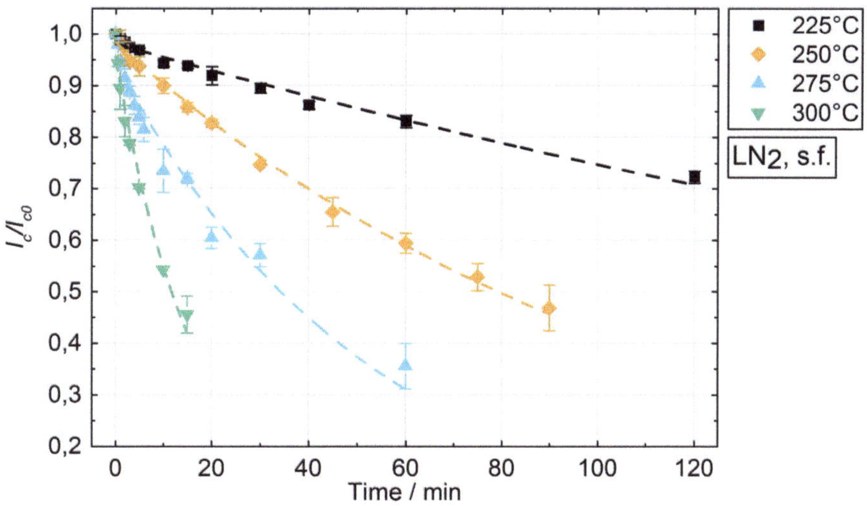

Figure 7.6: Overview of time dependent critical current degradation for SPI tapes

critical temperature as an indicator for a shift in the oxygen content of the *RE*BCO crystal was measured.

Figure 7.8 shows the electrical resistance as a function of temperature. One sample was treated for 1 hour at 225°C and one remained untreated. Table 7.3 summarizes the critical temperatures of the sample. The critical temperature $T_{c,10}$ was reduced by 1.5 K due to the heat treatment. In

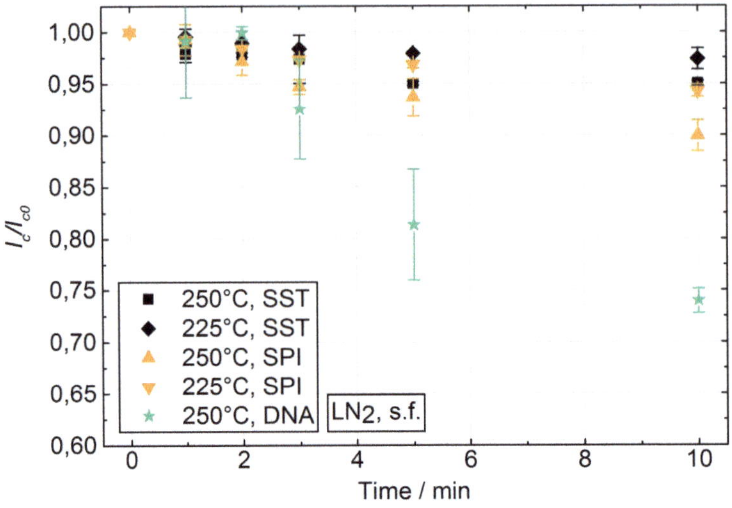

Figure 7.7: Critical current degradation of samples from three manufacturers at soldering temperature of 250°C.

addition to that reduction the transition width almost doubled from 0.2 K to 0.4 K. To check a connection between I_c degradation and possible oxygen loss, the exponential decays of I_c/I_{c0} at the different heating temperatures T_s were fitted (s. Fig. 7.6) . With the resulting decay constants τ the Arrhenius law was used and plotted ln τ over $1/T_s$ in Figure 7.9. An activation energy of 0.94 eV was found, which is in the order of activation energy of oxygen diffusion in *RE*BCO[Tis+99; Can+92].

Table 7.3: Summary of critical temperatures of untreated and heat treated SPI samples

	$T_{c,90}$ / K	$T_{c,10}$ / K	ΔT_c / K
Untreated	92.3	92.1	0.2
225 °C, 1 h	91.0	90.6	0.4
$\Delta T_{c,treat}$	1.3	1.5	

Assuming a linear temperature dependence between critical temperature and 77 K temperature (see. Figure 4.9) it is possible to roughly estimate the impact of T_c degradation on the I_c. The slope would be 5.53 A/K using $T_{c,10}$ with I_c ($T = T_c$) = 0 and I_c (T = 77 K) = 80.77 A. A decline of 1.5 K in critical temperature would therefore result in a decline of the critical current by 8.3 A. Relative to the untreated critical current this would be I_c/I_{c0} = 0.9 compared to the observed values in Figure 7.6 of I_c/I_{c0} = 0.83.

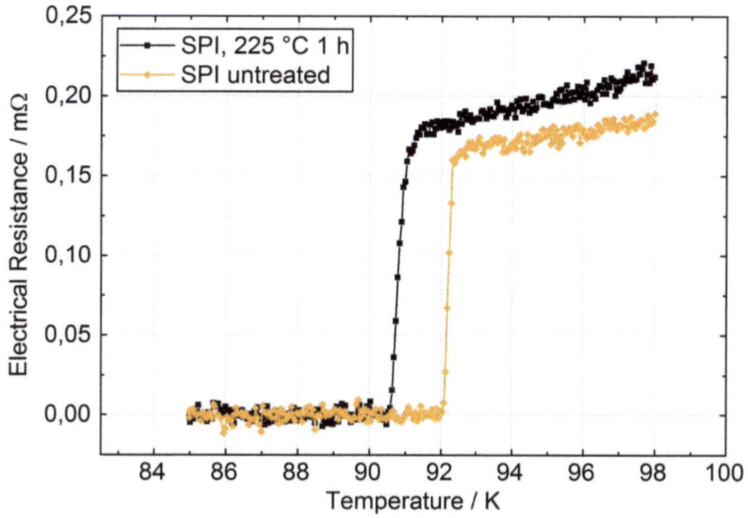

Figure 7.8: Critical temperature measurement of heat treated and untreated SPI tapes

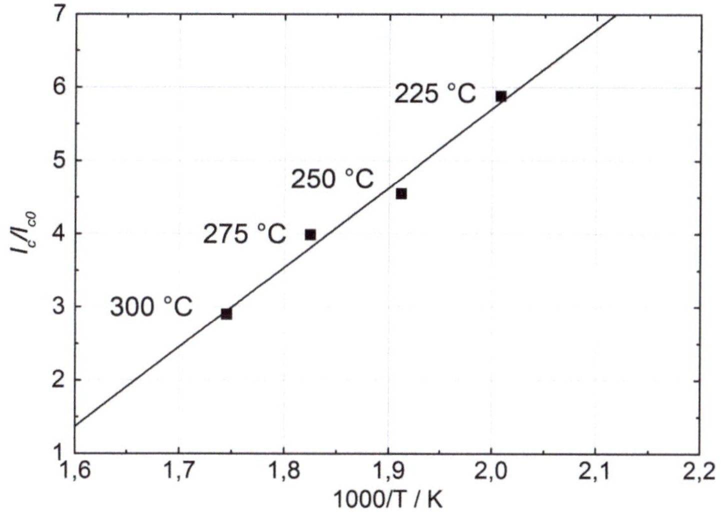

Figure 7.9: Arrhenius plot of the decay constants of the exponential decays in Figure 7.9

7.1.3 Conclusion

The degradation of *RE*BCO follows a predictable and repeatable pattern which, at least in part, appears to be caused by oxygen depletion of the *RE*BCO crystal. If oxygen diffusion is the only cause than it should be possible to reverse the critical current reduction by annealing tapes within an oxygen atmosphere. For this experiment the tapes would need to be only silver coated.

For the CroCo manufacturing it was determined that Sn-Ag solder might be a viable alternative to Sn-Pb with a soldering temperature of 250 °C. Measurable degradation appeared at 250°C for duration larger than 1 min. The soldering unit of the CroCo process is about 30 cm long which would translate to about 1 cm/s when one applies at 30 second safety margin.

7.2 CroCo manufacturing process and machine

The CroCo manufacturing process consists of six consecutive steps as shown in Figure 7.10. As a first step the tapes are arranged for example on spools as shown in Figure 7.11. The tapes are then pre-tinned which includes the application of a flux in order to remove contamination on tapes such as oxides and ensure optimal soldering of the tapes. Following the pre-tinning, the tapes are than brought into the characteristic cross shaped stack form and soldered together. An outer geometry can be applied such as the round solder matrix used in this work. The final step of the CroCo process is the jacketing of the CroCo with for example a round copper tube.

The first five steps of CroCo manufacturing can be performed on one machine while the final jacketing steps is performed separately. Figure 7.11 shows the CroCo manufacturing machine used in this project. The machine supports up to 60 spools with a tape capacity of 50 m to 100 m per spool depending on the thickness of the tapes. The to-be-soldered tapes are guided from the spools towards the soldering unit. Within the soldering unit the tapes form the cross-shape and leave the soldering unit embedded into a round solder matrix.

To ensure smooth forward motion, the machine uses a linear drive with a total length of 8 m that pulls the tapes through the soldering unit.

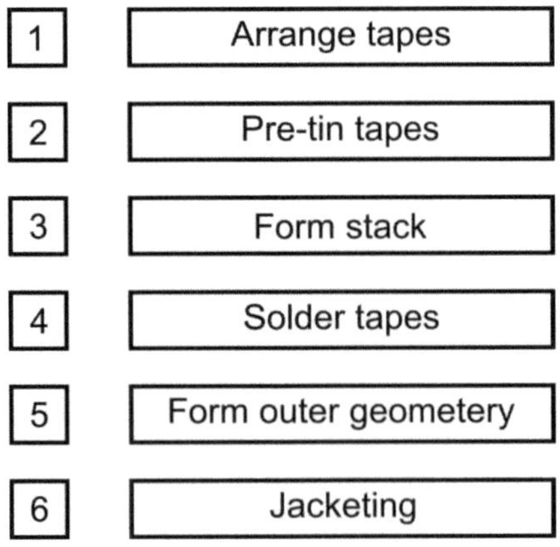

Figure 7.10: The six steps of the CroCo manufacturing process

Figure 7.11: CroCo manufacturing machine

In comparison to the proof of concept [Wol+16], the CroCo process has now integrated the outer geometry forming directly into the CroCo process instead of being an external process. When the first CroCo were fabricated, tapes with in total 100 μm electroplated copper stabilizer were used which lead to additional cost and restricted the number of available suppliers. The

CroCo process now uses *RE*BCO tapes with in total only 20-40 µm of electroplated copper stabilizer. Copper tapes are laminated between the *RE*BCO tapes to compensate for the lower thickness of electroplated stabilizer material.

The manufacturing machine shown in Figure 7.11 was a new development based on the experiences of the proof-of-concept machine to accommodate the changes to the CroCo process. During the CroCo manufacturing, the machine went through several iterations of optimization with the final one being shown in Figure 7.11.

7.3 Manufacturer qualification

Six manufacturers could in principle supply 6 mm and 4 mm wide tapes with electroplated copper at the beginning of 2017: SuperPower, SuperOx, SuNAM, Deutsche Nanoschicht (DNA), Theva, Shanghai Superconductors Technology (SST). Each company, with the exception of SuperPower, was contacted to supply 6 mm and 4 mm wide tapes for CroCo production qualification tests.

The CroCo process was initially developed on SuperPower tapes and therefore no additional test for SuperPower tapes were deemed necessary. Theva and SuNAM were at the time unable to provide sufficiently long pieces. SuperOx tapes were not acquired due to being significantly more expensive at that time than tapes from STT or DNA. Table 7.4 lists the basic tape specifications of the *RE*BCO tapes used in the manufacturer qualification.

Table 7.4: Tape structure specification of the *RE*BCO tapes used in the manufacturer qualification

	DNA	SST
Total thickness	100 µm	95 µm
Copper layer thickness	40 µm	40 µm
Substrate	60 µm Ni5W	50 µm Hastelloy
Critical current, 77 K s.f.	30 A/mm_{width}	30 A/mm_{width}
Tensile strength	150 MPa	600 MPa
Min Bending Radius	15 mm	7.5 mm

The qualification test was to manufacture a 2 m long CroCos consisting of 4 *RE*BCO and 16 copper tapes of 6 mm width as well as 2 *RE*BCO tapes and 8 copper tapes of 4 mm width. The copper tape thickness is 150 µm. The eutectic Sn63Pb37 (T_m = 183 °C) was used as solder with an appropriate flux.

For critical current measurements, the CroCos were cut into approximately 80 cm long pieces and the critical current was determined over a length of approx. 35 cm under LN_2, s.f conditions.

Figure 7.12 shows the *E*-*I* characteristic of the manufactured CroCos. Table 7.5 summarizes the measurement results and provides an expected critical current based on a FEM critical current simulation following the method described in section 6.3. The input for the simulation was the specified minimal critical current of the tapes and their angular critical current dependencies and therefore represent the minimum value the CroCo should achieve (see Section 4.3.2 and Table 4.3).

Table 7.5: Overview of the measurement and simulation results of the qualification CroCos. The expected critical current calculation is based on the minimal specified critical current of the acquired tapes.

Sample	I_c/ A	n	Expected I_c / A
SST	916	23.4	831
DNA 1	412	5.9	861
DNA 2	633	8.8	861
DNA 3	858	8.3	861

In total three CroCos containing tapes of DNA were manufactured. The first and second DNA CroCo (s. Fig. 7.12) were manufactured under the same conditions as the SST CroCo. The DNA 1 CroCo shows a significantly reduced critical current and n value compared to the SST CroCo. As this measurement represents only one data point the experiment was repeated without systematic changes. The second DNA CroCo shows an improvement of the critical current and n value but still below the expected critical current. Even though no systematic changes were performed, unconscious changes in for example the handling of the *RE*BCO tapes can not be excluded.

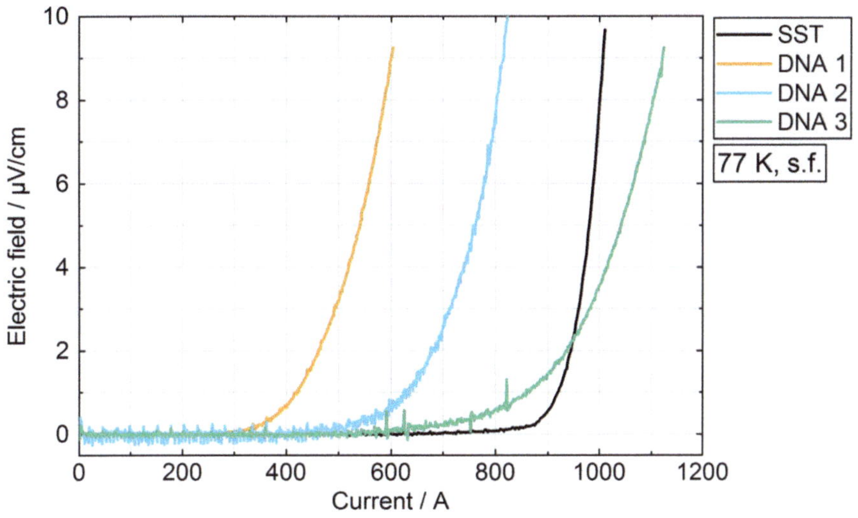

Figure 7.12: E-I characteristics at 77 K, s.f. during preliminary production tests for manufacturers selection

The observed critical current increase could be due to an improvement in the performance of the individual tapes. To investigate this, 15 cm long pieces were cut out from the spool at various distances and their critical current was measured in LN_2, s.f. conditions. In Figure 7.13 it can be seen that the critical current of the 4 mm wide tapes slightly increases along the tape but remained at 12.5 m still below the specified 120 A. For the 6 mm wide no such trend was observable however the sample size was small. The improvement of the critical current in the 4 mm tape is not sufficient to explain the improvement in critical current from DNA 1 to DNA 2. However it might indicate a warm-up phase during some of the DNA manufacturing processes for 4 mm tapes.

A cross section of a 4 mm DNA tape was prepared to investigate the tape structure for defects that could explain the critical current improvements. Figure 7.14 shows the edge region of a 4 mm wide DNA tape. In the top picture it can be seen that *RE*BCO layer delaminated from the substrate. The bottom picture shows that copper grew into the void, proving that the initial delamination did not occur during the cross section preparation or tape handling but rather during manufacturing.

The observed delamination could make the DNA tapes highly susceptible to mechanical stress which could result in the observed degradation of CroCos DNA 1 and 2.

A spring was inserted into the pulling apparatus for the final DNA CroCo (DNA 3) to reduce possible tensile stress spikes acting upon the superconductors.

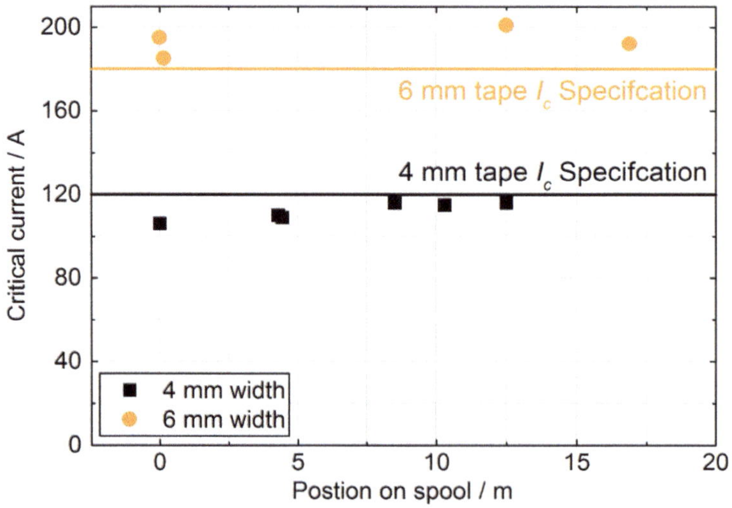

Figure 7.13: Critical current of 4 mm and 6 mm wide tapes of DNA at LN2, s.f. at various position along the delivery piece length

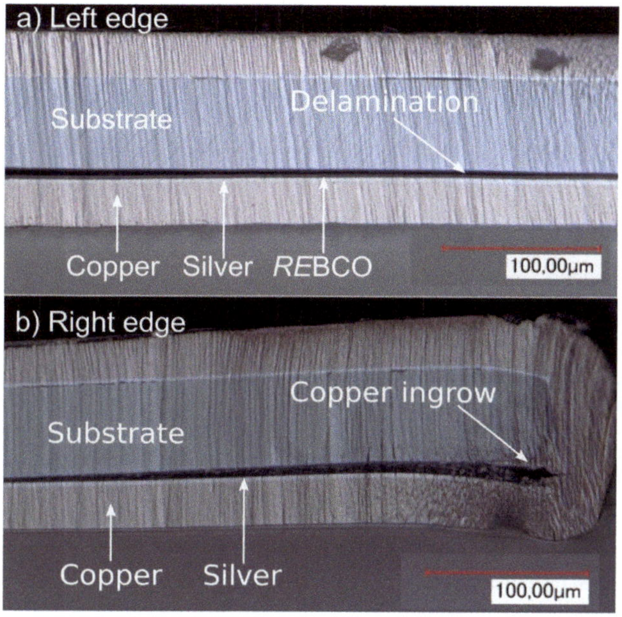

Figure 7.14: Cross section of the edge of a 4 mm DNA tapes a) left side edge region with delaminated copper and *RE*BCO layer b) right side edge region with delaminated copper as well as copper in-grow into delamination area . (Figures prepared by A. Jung)

With this measure DNA 3 achieved the expected critical current. However the low n value suggests that there is still degradation occurring during the manufacturing.

In conclusion the available DNA tapes were presently not suitable for the project. However DNA might become a viable choice for future projects if for example the delamniation would be resolved and additional measures to reduce tensile stress during fabrication are introduced.

Finally, out of the six manufacturer that could in principle supply the *RE*BCO tapes only SST and SuperPower were viable *RE*BCO tape manufacturers for CroCo production at the time. Out of these two options SST offered the cost effective option and was consequently chosen as supplier. The tape specification were identical with the specification listed in Table 7.4 expect for the stabilizer thickness. In order to save cost a thinner stabilizer of in total 20 μm was used.

The superconductors arrived in three batches spread over 4 months. Random samples were taken from the superconductor batches upon arrival and their critical current was measured. The measurements matched the critical currents data provided by the supplier. In addition, the thickness of the *RE*BCO tapes was measured with a micrometer gauge. Two measurements were performed, one across the tape and one along the center of the tapes. Figure 7.15 shows these measurements for two samples along the samples length. The distance between the positions is roughly 5 m. The grey area represents the expected tape thickness according to specifications. The tape thickness at the edges achieves the specifications however in the center

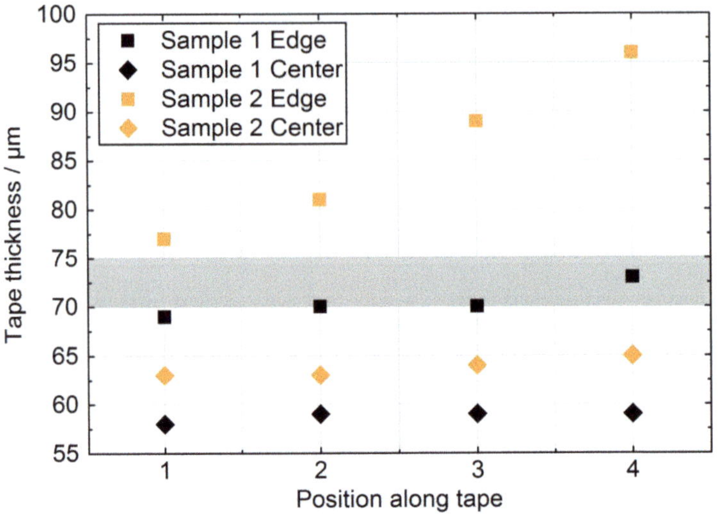

Figure 7.15: Measured tape thicknesses of two SST samples at various positions along the sample length

the tape thickness is significantly below the expected value with as little as 58 µm measured. This phenomenon is often referred to as *dog boning*.

Resolving this issue with the supplier would have incurred several months of delay to the project. It was previously decided that the majority of the required stabilizer would be provided by copper tapes. Therefore the missing electroplated stabilizer was not deemed critical and the project continued without exchanging the *RE*BCO tapes.

7.4 CroCo strand manufacturing

7.4.1 Preliminary CroCo manufacturing tests

The deployment of a new CroCo manufacturing machine as well as the integration of the outer geometry forming into the CroCo process necessitated preliminary tests to establish basic production parameters. The first tests was to manufacture CroCos consisting only of copper tapes in order to save superconductor. The CroCos in this step consisted of 20 x 200 µm thick 6 mm tapes and in total 10 x 200 µm thick 4 mm tapes. In the following these CroCos will be referred to as 'Pure copper CroCos'.

The second step was to substitute 4 copper tapes with *RE*BCO tapes in order to check for critical current degradation during the manufacturing. In the following these CroCos will be referred to as 'Partially filled CroCos'

The chosen criteria to move from pure copper to partially filled (with superconductors) CroCos was the surface smoothness of the CroCo, the dimensions of the CroCo, and how smooth the process runs. This last criteria describes how often operators had to intervene to ensure the desired results.

One of the easily controllable production parameters is the pull speed of the drive train. During the pure copper tests it was established that a pull speed of about 40 cm/min resulted in a smooth round solder matrix surface. Higher speeds resulted in defects in the surface of the solder matrix that appeared at irregular intervals (see Figure 7.16).

The substitution of individual copper tapes with *RE*BCO tapes had no significant impact on the surface quality of the solder matrix or the behavior of the process in general. An approximately 80 cm long sample was prepared and the critical current was measured over a distance of 40 cm. Figure 7.17 shows the critical current measurement of the partially filled CroCo. The n value is 16 which is low but acceptable. The critical current of CroCo is well above the simulated critical current, most likely due to better than specified performance of the individual tapes.

The results of the partially filled CroCos were sufficient to move towards a first test of fully packed CroCo.

7.4.2 CroCo manufacturing

The target CroCo composition in this project was 22 6 mm wide *RE*BCO tapes in a sandwich arrangement around 11 200 µm thick 6 mm wide copper tapes. The same sandwich geometry was adopted for the 10 4 mm wide *RE*BCO tapes around 4 200 µm thick copper tapes with an additional 2 100 µm thick copper tape for the outer 2 *RE*BCO tapes. The *RE*BCO layers of the tapes were arranged to face the copper tapes. This is schematically shown in Figure 7.18.

The first test with a fully stacked CroCo revealed a different behavior from previous tests. Surface defects as shown in Figure 7.16 appeared in regular 30-40 cm intervals. In addition the operators had to intervene several times for the process to progress. It took several iterations of optimization to improve the process to a level that no operator intervention required. It was not

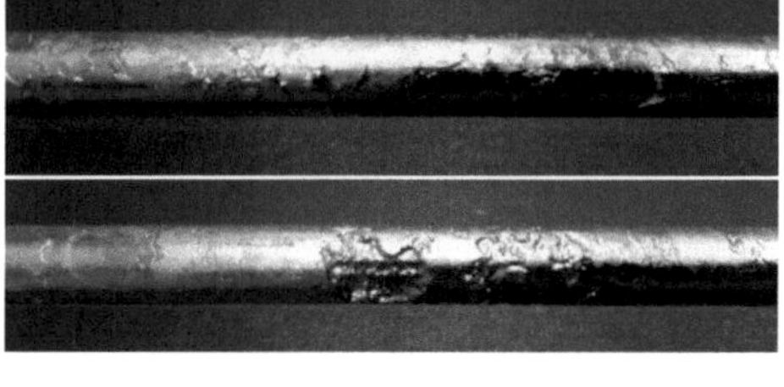

Figure 7.16: Surface of pure cropper CroCos. Top picture shows the desired smooth surface. Bottom picture shows the defects encountered at higher draw speeds

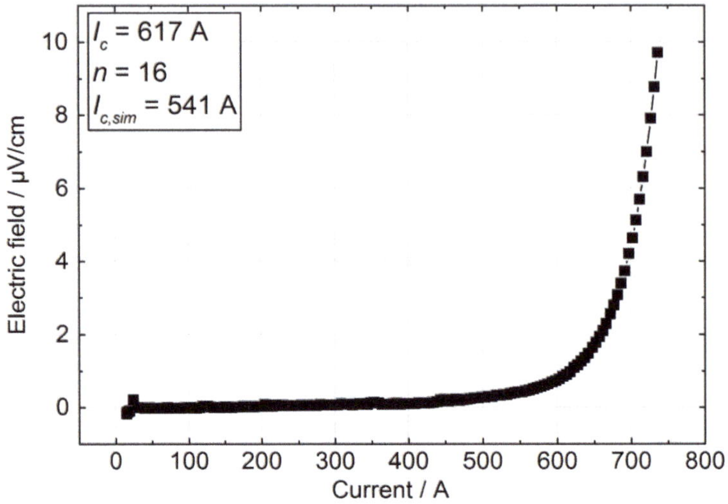

Figure 7.17: Critical current measurement of a partially filled CroCo with two 6 mm wide REBCO tapes and two 4 mm wide *RE*BCO tapes added

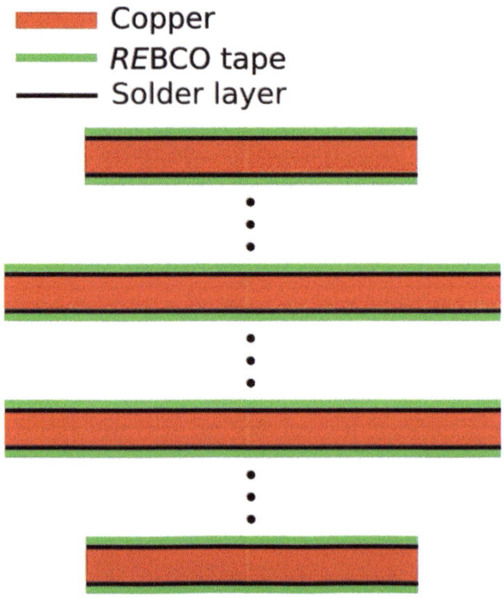

Figure 7.18: Schematic of the targeted CroCo structure. The *RE*BCO tapes are arranged in a sandwich structure around a relatively thick copper tape. The *RE*BCO layer of the *RE*BCO tape is facing towards the copper tape. Stacking multiple of these sandwiches on top of each other creates the desired CroCo structure

possible to fully reproduce the system behavior with copper tapes and therefore it was necessary to optimize the process with fully stacked CroCos.

Table 7.6 lists the length, manufacturer and geometry of each CroCo produced. The first three CroCo were produced in shorter length of approx. 4 m and then cut to the required 3.6 m. The remaining CroCos were produced in 8 m pieces and the 3.6 m were cut out of them. The second half of CroCo 6 was cut into shorter pieces and used in a different project.

Table 7.6: Summary of the produced length, used manufacturer and geometry of each CroCo

CroCo	Manufacturer	Length	Geometry			
			Copper tapes		*RE*BCO tapes	
			6 mm	4 mm	6 mm	4 mm
1	STT	4	$11 \cdot 0.2\,\mu m$	$8 \cdot 0.2\,\mu m$ $+2 \cdot 0.1\,\mu m$	22	10
2	STT	4				
3	STT	4				
4 & 5	STT	8				
6	STT	8				
7 & 8	STT	8				
9 & 10	STT	8				
11 & 12	SPI	8	-	-	22	10

7.4.3 CroCo residual production strain

A CroCo is a composite material with three major components: *RE*BCO tapes, copper tapes, and solder. The solder solidifies during the manufacturing of a CroCo effectively connecting the *RE*BCO tapes with the copper tapes. Strain and stress will occur within the CroCo as it cools down due to the different coefficient of thermal expansion (CTE) of its components.

Table 7.7 lists the CTE, young's modulus E, and offset yield strength R_{p02} of the components at 77 K. The *RE*BCO tape has the smallest CTE, largely defined by the Hastelloy substrate, which means that it will be compressed by the other materials during cool down. The critical current of a *RE*BCO tape is dependent on the strain (see section 4.3.4) and therefore it is important to investigate the magnitude of the *RE*BCO tape compression within a CroCo at operating temperatures.

The individual thermal strain of a component ε_i during cool down to a temperature T is given in [OMH16] as

$$\varepsilon_i = \int_{T_m}^{T} \alpha_i - \alpha_{CroCo} \mathrm{d}T \tag{7.1}$$

Table 7.7: Coefficient of thermal expansion (CTE), Young modulus E and offset yield strength R_{p02} of CroCo composite materials at 77 K

Material	CTE / 1/K	E (77 K)	R_{p02} (77 K)	References
*RE*BCO tape	$11 \cdot 10^{-6}$	121 GPa	540 MPa	[LCZ08; DS13; Cli+06]
Copper tape	$14 \cdot 10^{-6}$	116 GPa	488 MPa	[Mit05; Deu]
Sn63Pb37	$24 \cdot 10^{-6}$	50 GPa	90 MPa	[Fin+08; BBW14]

with α_i being the component CTE, α_{CroCo} being the CroCo CTE and T_m= 183 °C the solidus temperature of the Sn-Pb solder. Within the CroCo, the stresses which are caused by the contraction during cool down have to be in an equilibrium

$$\sum_i a_i \sigma_i = 0 \tag{7.2}$$

with a_i being the component volume fraction and σ_i the component stress. The stress-strain relationship needs to be defined to solve the force equilibrium.

Tensile tests were performed at 77 K in order to determine the stress-strain relationship of each of the CroCo component materials. The tests were performed in the Cryomak facility on the MTS 100 for the *RE*BCO and copper tapes and on the ATLAS test rig for the solder [SWB15]. The *RE*BCO and copper samples were cut from the spools with which the CroCos were produced. The *RE*BCO tape had a cross section 0.326 mm² while the copper had 1.194 mm². The solder samples were cut out of a 4 mm thick solder sheet and had a width of about 4.8 mm. The solder sheet was created by a fast cool down below the melting point in order to emulate the conditions during the CroCo manufacturing. Two samples were measured for each material and the averaged results are shown in Figure 7.19.

In Figure 7.19, the copper as well as the *RE*BCO tape show a large linear-elastic area up to a strain of about 0.3 % which can be modeled with the simplified Hooke's law

$$\sigma_i = E_i \varepsilon_i \tag{7.3}$$

The linear-elastic area of the solder is much smaller, instead an plastic deformations sets in relatively early. The elastic-plastic deformation curve can be fitted with an exponential function such as

$$\sigma_i = x_0 + x_1 \cdot \left(1 - \exp\left(\frac{\varepsilon_i}{x_2}\right)\right) + x_3 \cdot \left(1 - \exp\left(\frac{\varepsilon_i}{x_4}\right)\right) \tag{7.4}$$

A least-square fit over the data shown in Figure 7.19 determines the constants as x_0 = -4.234 MPa, x_1 = 67.73 MPa, x_2 = 0.011, x_3 = 82.24 MPa, x_4 = 0.002 for the solder.

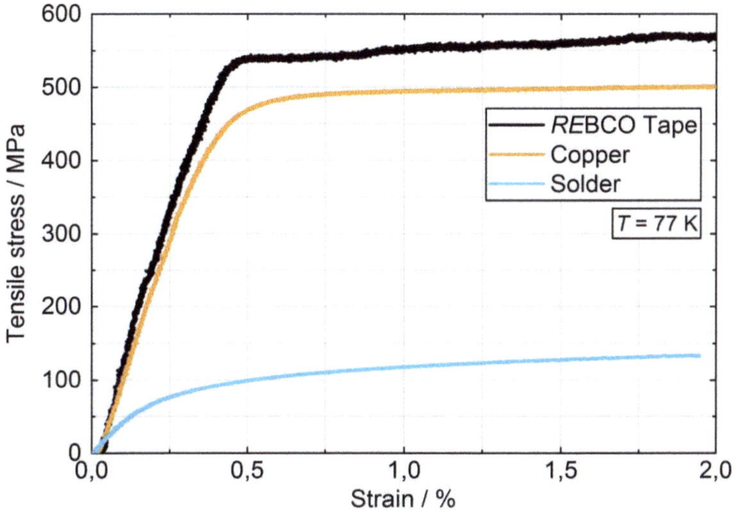

Figure 7.19: Stress-Strain curve of hardened copper and Sn63Pb37 solder at 77 K

Inserting equations 7.1, 7.3 and 7.4 into 7.2 allows to solve for the thermal expansion coefficient of the CroCo α_{CroCo}. Following that, the individual strains and stresses in each component can be calculated with equations 7.1, 7.3 and 7.4.

Figure 7.20 shows the residual strain of the CroCo component materials as a function of temperature between the solder solidus temperature of $T_{\mathrm{m}} = 456$ K and the operating temperature of $T_{\mathrm{op}} = 77$ K. The solder experiences the highest tensile strain due to being the softest material while copper remains almost strain free. The *RE*BCO tape experiences compressive strain of up to 0.09 % at 77 K.

The *RE*BCO layer within the *RE*BCO tape is also subjugated to residual strains during the manufacturing [Diz+14; OMH16] which need to be considered when determining the strain status of the *RE*BCO layer. In general, the increased compressive strain is positive since it counter acts tensile strain during for example bending and due to the fact that compressive strain does not degrade the critical current as fast as tensile strain (see section 4.3.4).

7.4.4 CroCo jacket

In general, a jacket improves the short circuit resilience of a CroCo and adds a protective layer to it. Therefore, the application of a copper tube jacket was investigated. Such jackets were previously realized by rotary swagging a copper tube onto the cross shaped CroCo and manually filling the gaps with solder [Wol+16]. To test if it is possible to rotary swag a copper tube on to the round solder matrix two 60 cm long fully stacked samples were prepared. The diameter of

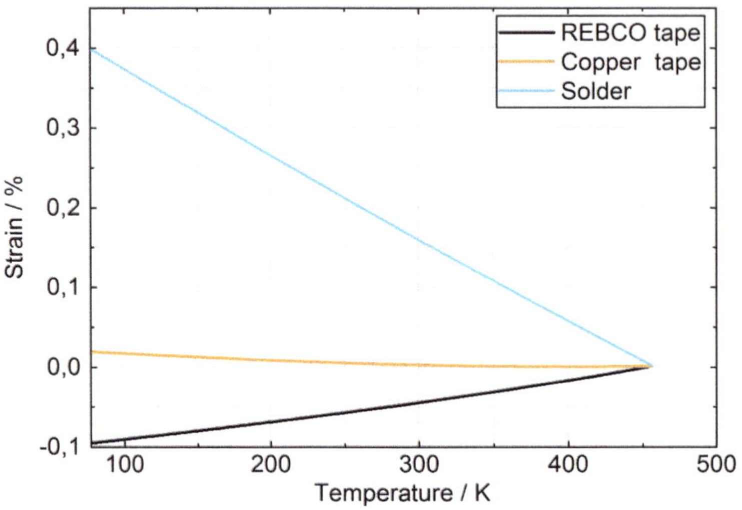

Figure 7.20: Residual strain of CroCo component materials between the solder solidus temperature of T_m = 456 K and the operating temperature of T_{op} = 77 K

the CroCo core was measured in the designated rotary swagging area every centimeter in two perpendicular directions. The diameter varied between 7.56 mm and 7.99 mm.

Figure 7.21 shows sample 1 prepared for a critical current measurement in LN_2, s.f. conditions. The rotary swagged area was approximately 20 cm and the distance between the voltage taps was 22 cm. The initial outer diameter of the copper tube was 10 mm with a wall thickness of 0.5 mm.

The rotary swagging reduces the outer diameter of the copper tube in steps of approximately 0.15 mm. Figure 7.22 shows the degradation of the samples over the outer diameter of the copper tube. At an outer diameter of 9.15 mm the copper tube could be moved freely along the CroCo core for both samples. At 9 mm the copper tube could still be rotated by hand around the copper tube for sample 1. At 8.88 mm outer diameter the copper tube could not be moved by hand but sample 1 was fully degraded. The measurements were discontinued for sample 2 because the degradation was already out of acceptable bounds while the copper tube could still be moved by hand.

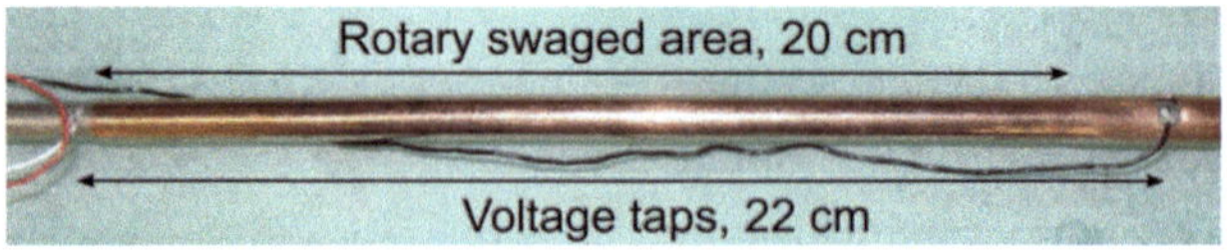

Figure 7.21: Sample 1 prepared for critical current measurement with rotary swagged area

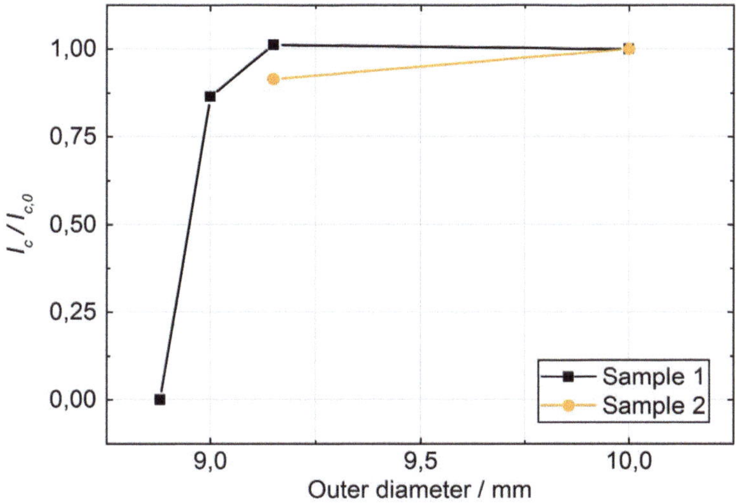

Figure 7.22: Degradation of fully stacked CroCos with declining outer diameter of a copper tube after rotary swagging

This behavior could be explained by the high variation in the outer diameter of the CroCo core with almost 0.5 mm between the minimum and maximum outer diameter. At the largest diameter the the copper tube would be pressed against to CroCo and cause mechanical degradation while a gap remains along the rest of the CroCo.

From Figure 7.22 it can be concluded that jacketing with the existing rotary swagging machine is not a viable solution for the round solder matrix CroCo in its current iteration. This is due to the fact that degradation occurred while the copper tube was still movable by hand. Suggesting that the contact between copper tube and CroCo core is limited to very few points. This would than defeat the purpose of the jacketing (mechanical, thermal and electrical stabilization). If the jacketing is not LN_2 tight it could even harm the thermal stability of the CroCo in case of a quench. During a quench the LN_2 would evaporate and be contained by the jacketing effectively creating an insulating gas bubble around the CroCo core.

Figure 7.23 shows a cross section of sample 1 after it was fully degraded. It can be seen that the 6 mm wide *RE*BCO tapes are slightly vertically out of position. Non of the *RE*BCO tapes appear to bend or otherwise significantly altered that could explain the full critical current degradation. This would indicate micro fracture directly within the superconducting layer as a cause for the degradation.

For this reason no jacketing was applied to the CroCos in this project. This was possible since the jacketing only acted as thermal stabilizer under extreme conditions which were not scheduled to be tested within the frame of this work.

Figure 7.23: Cross section of sample 1 after its critical current fully degraded.

The rotary swagging approach is driven by CroCo development for fusion applications where a seamless jacket offers advantages due to large Lorentz forces acting upon the CroCo which could lead to cracks along the seam. For power applications non-seamless solutions such a wrapping could be a scalable solution for long length CroCos. However this investigation was out of the scope of this work.

8 35 kA *RE*BCO DC cable test

8.1 Single CroCo characterization

The electrical characterization of the CroCo and its terminals consists of a critical current measurement with a constant ramp rate and a constant current experiment in order to verify steady state operation at elevated currents. Microscopic cross sections were prepared to assess the quality of the CroCo shape in regards to the roundness of the solder matrix and the alignment of *RE*BCO tapes. Finally, the electrical characterization allows for a validation of the current capacity of the assembled demonstrator.

8.1.1 Electric CroCo characterization

Figure 8.1 shows a schematic side view (end section) and cross section of the measurement setup for the single CroCo characterization.The CroCos and their terminations are mounted with glass fiber reinforced plastic (GFRP) holders on a 5 m long aluminum profile. The aluminum profile is then placed on GFRP blocks into an 5.5 m long open stainless steel tank that is insulated with 50 mm of Styrodur on each side. After connecting the termination's cable shoe with two 300 mm² of copper cables to the source, the stainless steel tank is filled with liquid nitrogen until the cable shoe is submerged. The source consists of 4 Keysight 6680A DC power supply systems with a total current capacity of approximately 3500 A. The data acquisition is described in section 8.2.4.

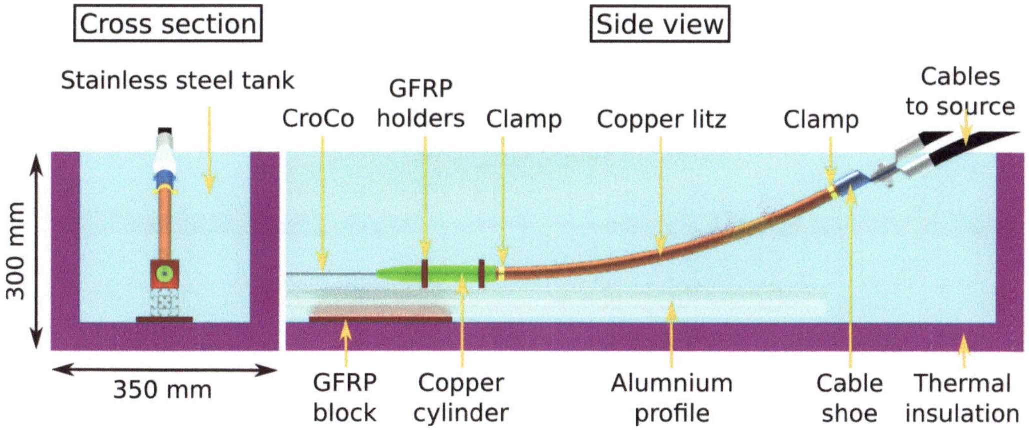

Figure 8.1: Cross sectional and side view of the experimental setup for single CroCo characterization measurements in LN_2

Figure 8.2 displays a schematic (not to scale) of a CroCo with its terminations and the voltages that were measured during single CroCo characterization. At both ends, the CroCo is soldered over a length 10 cm into a grooved copper cylinder (CC) with a low melting solder. The copper cylinder in turn is connected through 300 mm² of copper litz (CL) to a cable shoe (CS) to which additional cables can than be connected as shown in the experimental setup (Figure 8.1). The connection of the copper litz wire to the neighboring components is realized through soldering with eutectic Sn-Pb solder. Soldering posts are added to each of the components to which voltage taps can be connected as for example schematically shown in Figure 8.2.

The cable shoe solder posts are screwed to the center of the cable shoe's inclined surface while the copper litz solder posts are soldered to a copper clamp surrounding the copper litz. The clamps are mounted on the copper litz with a distance of 10 mm to cable shoe and copper cylinder respectively. The solder post within the copper cylinder is placed at the end of the groove into which the CroCo is soldered. No solder posts were placed on the CroCo itself, instead voltage taps are soldered directly to the CroCo's solder matrix surface with a lower melting solder.

The purpose of the cable shoe to copper litz and copper litz to copper cylinder measurements is to assess the quality of the soldered connections between copper litz wire, cable shoe and copper cylinder respectively. Thus enabling intervention in case the resistances of the connections are not within acceptable bounds relative to the other CroCos.

By measuring the CL resistance it is possible to use the CL as a shunt in the demonstrator assembly and thus enabling a direct measurement of the current distribution across the CroCos.

The contact resistance of the CroCo (CC-CroCo) is measured to provide statistical data about the spread of the resistances that can achieved with such a connection. The voltage tap on the CroCo is placed with a distance of 10 cm from the edge of the copper cylinder. Directly along the CroCo two voltage measurements are performed by using voltage taps with a distance of 2 m and 2.8 m respectively.

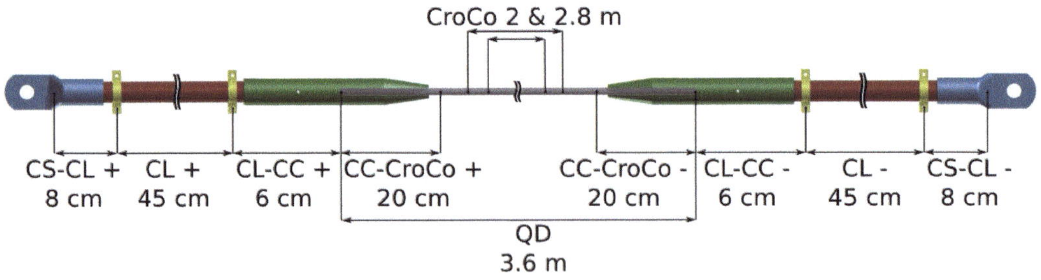

Figure 8.2: Single CroCo characterization cabling plan (not to scale)

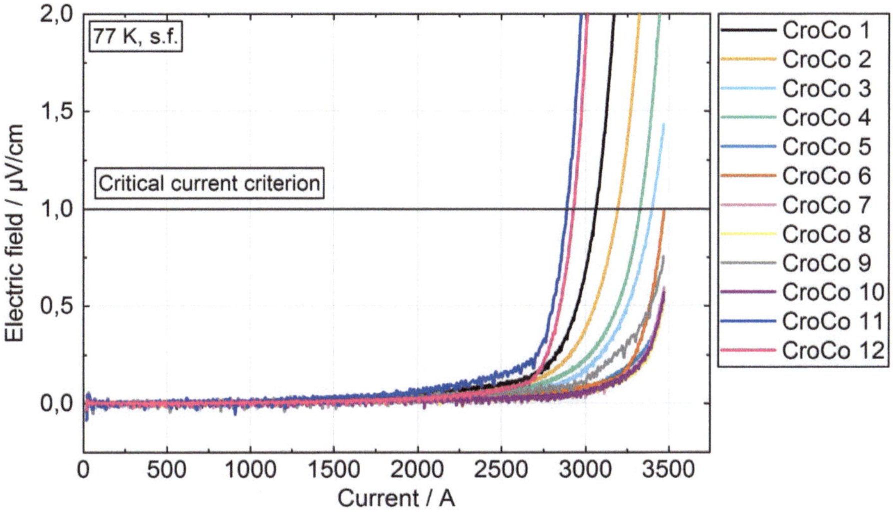

Figure 8.3: Electrical field as a function of current for all 12 CroCo over the 2.8 m voltage taps

The voltage along the entire length of the CroCo including soldered connections is measured and used as a guideline for setting up the quench protection threshold, which uses the same voltage taps.

Two types of experiments were performed during the single CroCo characterization. During the first one, the current was ramped up with a constant ramp rate of approximately 50 A/s until the quench detection triggered or the maximum current of the source (approx. 3.5 kA) was reached. For a constant ramp rate any inductive signals that are be picked up in the measurement remain constant and therefore can be compensated in the data analysis.

The second experiment was to ramp up to a constant current and to hold it for at least 3 min before ramping down. This was done to ensure that the CroCos were capable of carrying the expected current for a long time and to observe if the components enter a steady state.

CroCo critical current

Figure 8.3 shows the electrical field of all 12 CroCos as a function of the current of the 2.8 m voltage taps at 77 K. In general, a relatively large spread in the critical current can be observed with the lowest critical current being 2890 A and the highest in the area of 3680 A. Notably the lowest performing CroCo are the last two CroCo manufactured with SPI tapes. The critical current of the STT CroCos increased after the first four CroCos beyond the capabilities of the current source of approx. 3.5 kA.

Table 8.1 lists the measured and calculated critical currents of the CroCos. The measured critical current is determined through a power-law least square fit (see section 4.1) over a measurement with constant current ramp. Constant voltage offsets caused for example by inductance were removed prior to the fit. The details of critical current calculation are discussed in Section A.5.

The *n*-values listed in Table 8.1 vary between 10 and 25 even though some of the lower values for example from CroCo 5 might be due to a lack of data for a sufficiently accurate fit in the region. The critical current of CroCo 5 and CroCos 7 to 10 is above capabilities of the current source. Therefore the listed values of these CroCos need to be considered as an extrapolation rather than a measurement.

Table 8.1: Measured critical current and n-values of all CroCos for voltage taps with 2 m and 2.8 m distance at 77 K, s.f. and calculated critical current values based three different scenarios (see Section A.5)

	Measured				Calculated		
CroCo	I_c(2 m)	n (2 m)	I_c (2.8 m)	n (2.8 m)	Avg.	Max.	Min.
	A	-	A	-	A	A	A
1	3080	18	3060	19.4	3460	3194	3706
2	3210	18.3	3193	16.4	3512	3240	3938
3	3398	20.6	3410	17.4	3691	3237	3993
4	3437	18.5	3320	15.4	3743	3341	4079
5	3680	13.5	3699	12.4	3743	3341	4079
6	3482	24.5	3485	22.4	3600	3200	3903
7	3597	18.5	3596	19	3692	3296	3296
8	3618	18.2	3632	18.2	3692	3296	3296
9	3621	11	3607	11	3676	3287	3984
10	3607	18.1	3615	18.3	3676	3287	3984
11	2890	24	2890	24	-	-	2995
12	2928	24.9	2926	24.9	-	-	2995

Figure 8.4 shows the measured critical currents, and calculated critical currents as listed in Table 8.1. The critical current measurements across the two different distances are almost identical except for CroCo 4. The difference in critical current might indicate some damage to the CroCo in the outer areas of CroCo 4. It can be noted that the expected critical current of the first 3 CroCos is lower than the remaining CroCos and that the performance of the first 4 CroCos is better predicted by the simulation based on the minimum tape data. For CroCos 5 to 10 the critical current based on average tape data is more accurate. The improvement in critical current predictability based on average critical currents can be attributed to a learning curve and successive improvements in the manufacturing process.

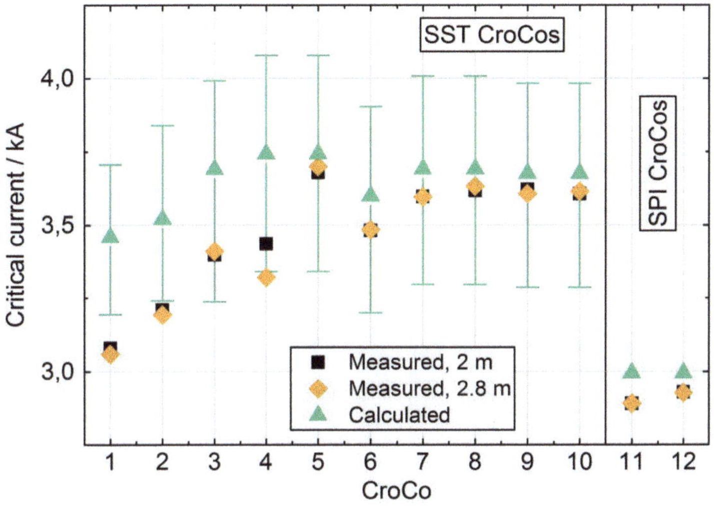

Figure 8.4: Critical currents of the all CroCos compared with simulated critical currents based on minimum, average, and maximum tape critical currents

Termination resistances

Figure 8.5 shows and Table 8.2 lists the termination resistance of the CroCos and of the components in the termination. The resistance was analyzed by two methods: a linear voltage-current fit between 500 A and 1500 A and an average at 100 A for at least 60 s. Since the results of the two methods do not show significant deviation only the first method is shown.

The major resistive parts are the copper litz (CL) with a resistance of about 3 $\mu\Omega$ per pole. The connections between copper litz and the cable shoe (CL-CS) are in the range of about 0.5 $\mu\Omega$ to 0.9 $\mu\Omega$ per pole as are the connections between copper litz and copper cylinder (CL-CC). The joints between the copper cylinder and the CroCo show the lowest resistance within the termination with 0.15 $\mu\Omega$ to 0.32 $\mu\Omega$.

The total resistance varies between 7.81 $\mu\Omega$ at CroCo 12 and 10.08 $\mu\Omega$ at CroCo 6. The high resistance of CroCo 6 is largely due to the high copper litz wire resistance at the minus pole.

It can be noted, that for the CroCo connection (CC-CroCo) no significant deviation between the two manufacturers SST (CroCo 1 to CroCo 10) and SPI (CroCo 11/12) can be observed. This result indicates that it is possible to produce low resistance joints with CroCos that utilize laminated copper instead of electroplated [WFP17].

The copper litz wires are used to estimate the current distribution within the cable demonstrator and therefore it is important to determine if a voltage drift occurs due to for example increased temperatures at high currents. Several constant current measurements were performed for each CroCo between 2 kA and 3.5 kA depending on the critical current of CroCo. Figure 8.6

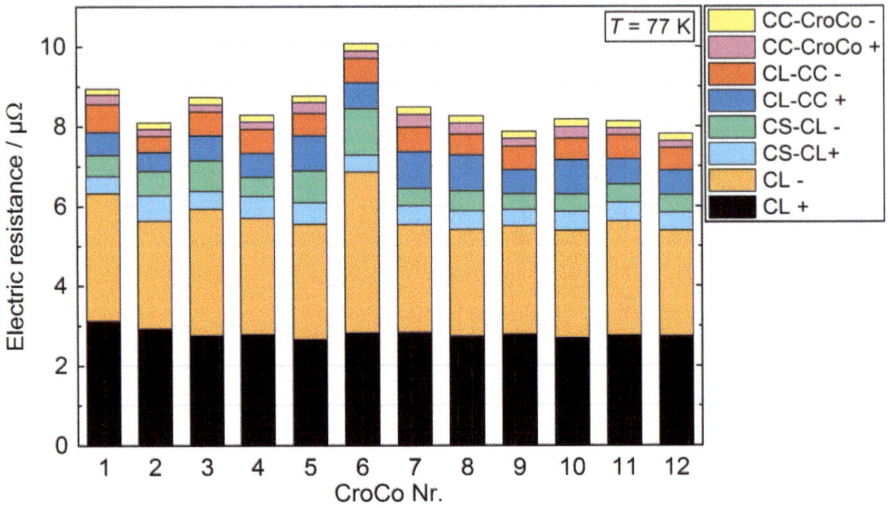

Figure 8.5: Termination resistances (s. Fig 8.2) of all CroCos at 77 K and determined between 500 A and 1500 A

Table 8.2: List of termination resistances (s. Fig 8.2) of all CroCos at 77 K and determined between 500 A and 1500 A

CroCo -	CL+	CL-	CS-CL+	CS-CL-	CL-CC+ \ µΩ	CL-CC-	CC-CroCo+	CC-CroCo-
1	3.12	3.21	0.42	0.53	0.57	0.7	0.25	0.15
2	2.93	2.71	0.64	0.62	0.47	0.4	0.18	0.16
3	2.76	3.17	0.45	0.77	0.63	0.59	0.18	0.18
4	2.79	2.92	0.55	0.48	0.6	0.6	0.18	0.18
5	2.67	2.89	0.54	0.8	0.88	0.56	0.28	0.16
6	2.82	4.04	0.42	1.17	0.65	0.61	0.19	0.19
7	2.84	2.71	0.47	0.44	0.92	0.61	0.32	0.19
8	2.74	2.68	0.47	0.51	0.89	0.53	0.27	0.18
9	2.79	2.72	0.41	0.4	0.6	0.59	0.19	0.18
10	2.7	2.7	0.47	0.43	0.87	0.54	0.28	0.19
11	2.77	2.86	0.47	0.45	0.63	0.6	0.17	0.18
12	2.75	2.65	0.45	0.44	0.61	0.57	0.17	0.18

shows a constant current measurement for CroCo 4. The current is ramped with about 50 A/s up to 3.5 kA where it remains constant for 3 min and then is ramped down with 100 A/s. A small voltage drift can be observed during the constant current period for the voltage across the copper litz, all others remain in steady state. The voltage increases by about 0.35 mV (CL +) and

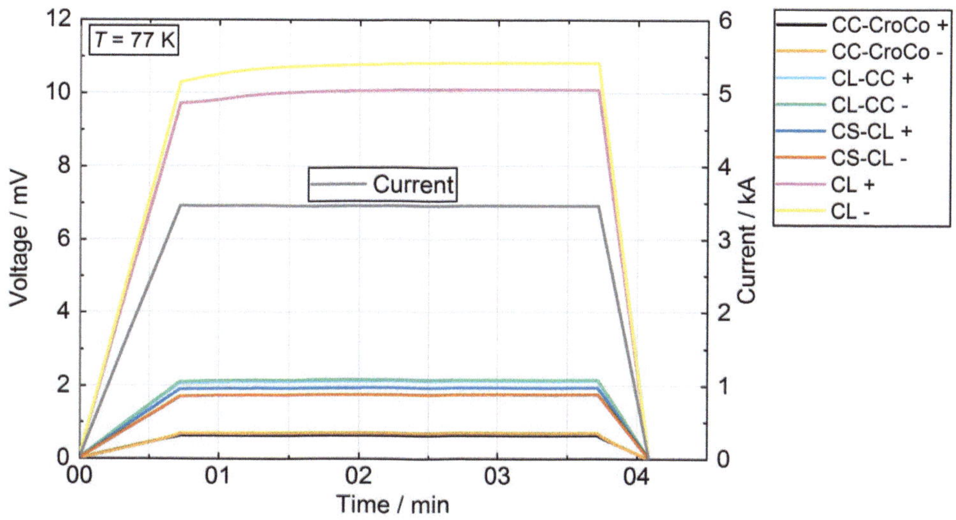

Figure 8.6: Voltage development of CroCo 4 during a 3 min constant current of about 3.5 kA

0.54 mV (CL -) and appear to enter a steady state after 3 min. The voltage drift relative to the total voltage is therefore in the region of 5 %.

It is assumed that voltage drift is due to a temperature increase at high currents which is a function of the heat transfer between copper wires and liquid nitrogen. The heat transfer in turn is largely impacted by the available surface area for the heat transfer. A tightly packed copper litz wire offers less surface area than a loosely packed one and therefore the steady state resistance at a certain current can vary.

The results of the electrical characterization are published in [Pre+19].

8.1.2 Microscopic characterization

Shown in Figure 8.7 are the cross sections of CroCo 1 to CroCo 8 and in Figure 8.8 the cross sections of CroCos 9 to CroCo 12. For each CroCo production run (see Table 7.6) a cross section was prepared and therefore multiple CroCos for example 4 & 5 share the same cross section.

In general, one can recognize the thicker copper tapes which are laminated between the thinner *RE*BCO tapes as described in section 7.4.2 for CroCo 1 to CroCo 10. In CroCo 11 and CroCo 12 the thicker electroplated stabilizer layers of the *RE*BCO tapes are visible which made it possible to forgo additional laminated copper.

In detail, CroCo 1 shows severe deformation of two *RE*BCO tapes that are completely outside the regular CroCo structure and several other tapes that are slightly out of position. The deformation of the out-of-position *RE*BCO tapes likely resulted in a strong critical current degradation thus contributing to the lower critical current of CroCo 1. Furthermore the solder matrix has an irregular oval shape.

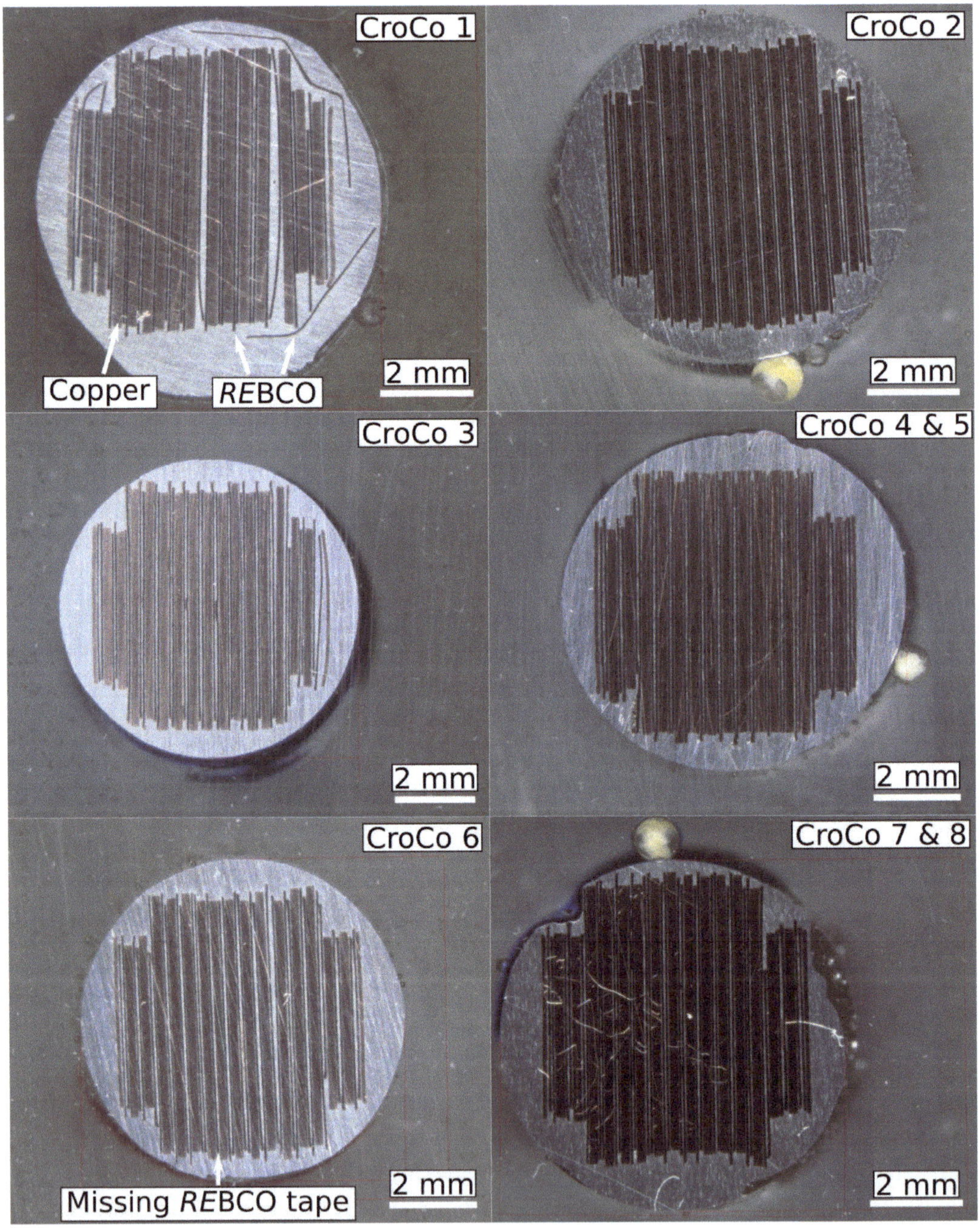

Figure 8.7: Microscopic cross section of CroCo 1 to CroCo 8

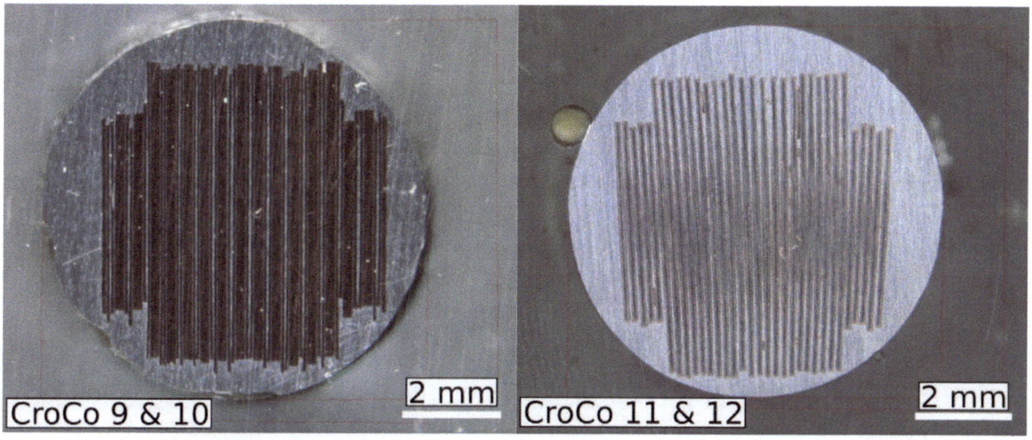

Figure 8.8: Microscopic cross section of CroCo 9 to CroCo12. The CroCos were produced in pairs and only on cross section was produced

Major improvements in terms of the CroCo structure integrity and solder matrix roundness can be observed from CroCo 1 to CroCo 2. The remaining small displacements of *RE*BCO tapes relative to their copper tapes are still within manufacturing tolerances and could likely be further reduced by decreasing the dog-boning of the *RE*BCO tapes.

In CroCo 3 one of the inner 4 mm tapes slipped into 6 mm structure due to wide tolerances in the stack forming process. This phenomenon reoccurs in CroCo 5 to CroCo 8.

In CroCo 6 one 6 mm wide *RE*BCO tape is missing which is due to a rupture during the manufacturing. The structure of the CroCo remains unaffected by rupture as it occurred early in the process.

The cross section for CroCo 7 and CroCo 8 is embedded in a solder matrix with a defect, as shown in Figure 7.16. Thus, the shape of the solder matrix shown in the figure is not representative for the rest of the CroCo length.

The last two CroCos (11 and 12) used tapes from SuperPower Inc with 100 μm electroplated copper stabilizer. Therefore no additional copper tapes were necessary as can be seen in Figure 8.8.

Figure 8.9 shows an example of solder layer thickness measurements that were performed optically on CroCo 6. The solder layer thickness usually varied between 5 μm and 40 μm with some outliers as for example seen in Figure 8.7, CroCo 3 on the right side.

Figure 8.9 also shows the cross section CroCo 6 with the missing *RE*BCO tape in detail. No variation in terms of solder layer thickness are observed due to the missing *RE*BCO tape.

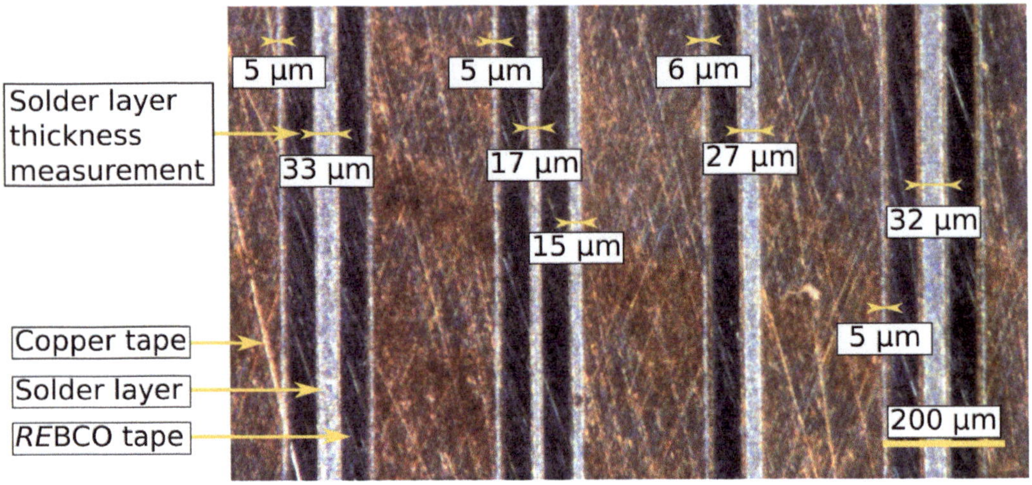

Figure 8.9: Solder layer thickness measurements on CroCo 10

8.1.3 Current distribution simulation

Based on the characterization of the CroCos and the individual termination resistance (Section 8.1.1) it is possible to calculate the expected current distribution in the demonstrator. The CroCos are connected in parallel and therefore the voltage drop is given by

$$U = (R_i + R_{end})I_i + E_c L \left(\frac{I_i}{I_{c,i}}\right)^{n_i} \tag{8.1}$$

with R_i being the sum of all terminal resistances measured (Table 8.2), $I_{c,i}$, and n_i being the CroCo specific properties (Table 8.1) and R_{end} an additional contact resistance between the terminations cable shoe and the copper block (see Fig. 8.14).

The current I_i flowing in each CroCo can be numerically calculated with equation 8.1 for a given voltage U. This has the advantage that each equation can be solved independently instead of a system of 12 highly non-linear equations as it would be the case for a given total current I.

Solving Equation 8.1 with the electrical properties determined in Section 8.1.1 (Tables 8.2 and 8.1) results in the current distribution as shown in Figure 8.10. Displayed is the current flowing in the individual CroCo as the function of total current. In general, the relation is largely linear due to the large termination resistance. CroCo 6 shows a distinctively lower current compared to the other CroCos due to its relatively high termination resistance. CroCo 11 and CroCo 12 are likely to be the limiting factor as they have relatively low termination resistance and the by far lowest critical current. This result in a deviation from the linear behavior at around 35 kA for both CroCo 11 and CroCo 12.

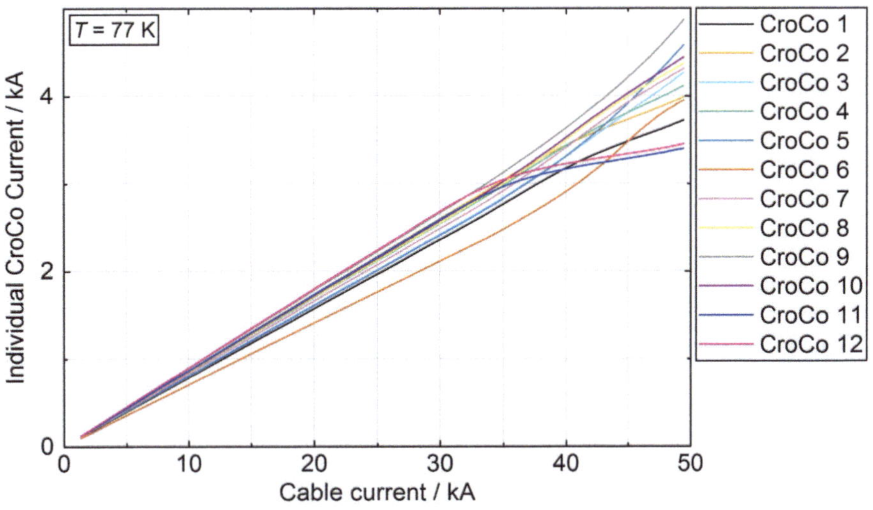

Figure 8.10: Individual CroCo current as a function of the total current flowing through the cable

For the simulation an R_{end} of 500 nΩ was estimated based on contact resistance measurements with thin copper sheets. Based on the experience of single CroCo measurements it was deemed safe to the set the quench limit to 3 mV. At 3 mV CroCo 11 and CroCo 12 are expected to be in the flux-flow regime above their critical current.

8.2 Demonstrator cable setup

8.2.1 Cryostat

Figure 8.11 shows a schematic of the cryostat built for this experiment. It is a double walled cryostat with inner dimensions of 5870 mm x 870 mm x 485 mm in length, widths, and height respectively. The insulation consists of 50 mm thick Styrodur. Five overlapping lids can be placed on top of the cryostat to reduce heat influx and ensure controlled removal of gaseous nitrogen. The liquid nitrogen enters the cryostat through the central lid and is pumped from the cryostat at both ends as it evaporates. The nitrogen is heated up through water baths to avoid damaging the pumps before being released to the environment. Care needs to be taken during operation to avoid an accumulation of liquid oxygen within the cryostat.

The cryostat is equipped with 9 Pt 100 temperature sensors which are placed in groups of three at the center of the cryostat, and at each end of the cryostat as shown in Figure 8.12. The vertical distance between the temperature sensors is 20 cm which results in heights measured from the bottom of 2 cm, 22 cm and 42 cm respectively. The Pt 100 are mounted on top of a small

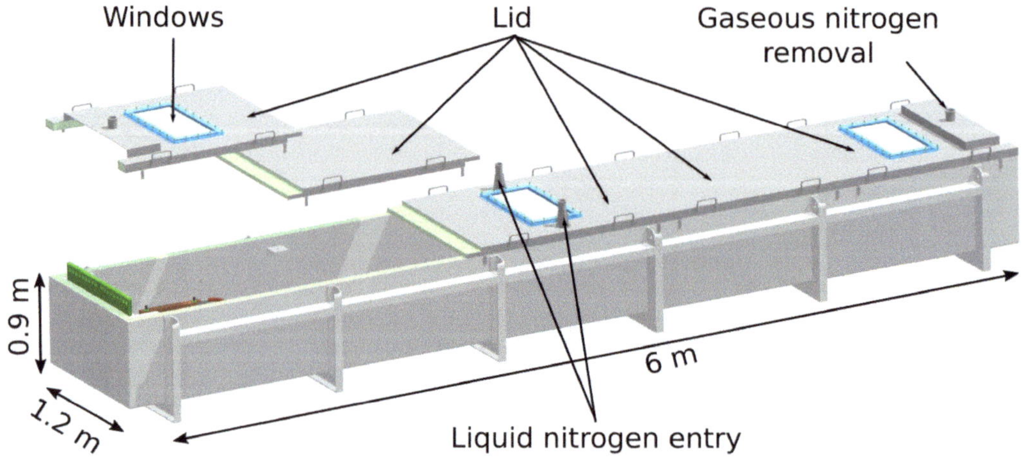

Figure 8.11: Liquid nitrogen cryostat used in the main experiment

copper plate which in turn is screwed to the surface of the inner cryostat wall. The purpose of the temperature sensors is to monitor the temperature gradients within the cryostat wall during cool down and thus helping to prevent damage to the cryostat through for example thermal expansion.

A pressure gradient will develop within the LN_2 as the cryostat is filled and therefore a temperature gradient could develop. The target fill height of cryostat during an experiment is $h = 35$ cm and thus the maximum pressure gradient within the LN_2 assuming an in-compressible fluid is

$$\Delta p_{LN2} = h \cdot \rho \cdot g = 0.35\,\mathrm{m} \cdot 802\,\mathrm{kg/m^3} 9.81\,\mathrm{m/s^2} = 2754\,\mathrm{Pa} \tag{8.2}$$

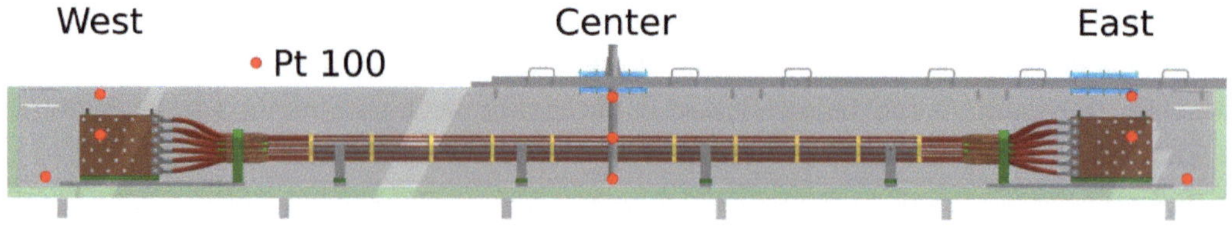

Figure 8.12: Location of PT 100 sensors within the cryostat

Assuming saturation properties the temperature gradient would be about $\Delta T = 0.13\,\mathrm{K}$ between the surface and the bottom of the cryostat [Lin97]. The pressure induced temperature gradient between the lowest and highest placed CroCo is about $\Delta T = 0.065\,\mathrm{K}$.

The heat leakage through the cryostat is estimated as

$$Q_{Styrodur} = \frac{\lambda A \Delta T}{s} = \frac{0.03\,\frac{\mathrm{m^2K}}{\mathrm{W}} 16.75\,\mathrm{m}^2\,(300\,\mathrm{K} - 77\,\mathrm{K})}{0.05\,\mathrm{m}} = 2241\,\mathrm{W} \tag{8.3}$$

the thermal conductivity of Styrodur λ was linearly extrapolated to 77 K based on manufacturer information. The leakage through the current leads at $I = 0$ A is equal to

$$Q_{Copper} = \frac{A \cdot N}{L} \int \lambda dT = \frac{0.0003\,\mathrm{m}^2 \cdot 36}{0.5\,\mathrm{m}} \cdot 90000\,\frac{\mathrm{W}}{\mathrm{m}} = 1944\,\mathrm{W} \tag{8.4}$$

with λ being the thermal conductivity of copper. During an experiment there will be additional losses due to ohmic losses in the resistive parts of the experiment. At a current of 40 kA the resistive losses amount to about 1500 W. In total the expected losses are therefore about 5700 W.

For one experiment a consumption of about 5000 l of liquid nitrogen was estimated. This split into 868 l for cooling the heat capacity of all major cold components, 1766 l for the required liquid nitrogen level within the cryostat, and 2400 l due to cryostat losses assuming an experiment duration of 24 h. The coolant is supplied directly from the liquid nitrogen storage tank through a vacuum insulated supply line coupled with manually controlled valves. In addition, transportable liquid nitrogen tanks can be connected to the cryostat to increase the fill rate.

Figure 8.13 shows the closed cryostat during a cool down test. Clearly visible are the water bath heaters at the ends of the cryostat through which the gaseous nitrogen is pumped. Also shown are two nitrogen supply lines, one with liquid nitrogen and one with room temperature nitrogen supply line which is used to remove air and therefore humidity from the cryostat prior to the cool down.

On the left side, the copper cables leading out of the cryostat and connecting to a copper block are visible. Connected on the other side of the copper block are water cooled cables that lead to the power source. The copper block itself needs to be heated to avoid freezing the water during cool-downs. The significant heat loss (sec. 8.2.1) by the copper cables leading out of the cryostat is further illustrated by the snow cap developing around them.

8.2.2 35 kA cable demonstrator

In Figure 8.14 the design schematic of the 35 kA cable demonstrator is shown in a top and side view as well as two cross sectional views at different positions. The 12 CroCos are mounted with braces upon an aluminum core with a diameter of 110 mm and 3 m length. Profile B in Figure 8.14 shows the location of the the previously characterized CroCos around the aluminum core. The copper cylinders at the end of the CroCos are placed into GFRP holders with a GFRP core

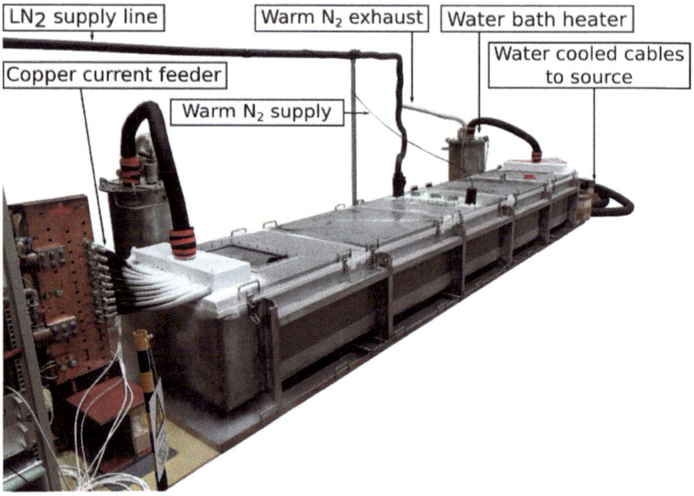

Figure 8.13: Cryostat with water bath heaters during cool down

in the center that allows for axial movement but restrict any radial movement. This helps to minimize any strain occurring due to temperature gradients during cool-down. The termination cable shoe's are screwed with a defined momentum to large copper blocks placed within the cryostat. Up to three copper blocks per side can fit into the cryostat therefore making it possible to connect CroCos in series. This was implemented to allow for experiments exceeding the current limitation of the power source in the future. The copper blocks are connected to the power source through several of copper cables.

The challenge of this termination method is that the current distribution across the CroCos is largely determined by the resistance of the terminals and the soldering connections. Current redistribution will be very limited if a CroCo should quench. This necessitates a fast quench detection for each CroCo in order to prevent damage during quench or fault currents. Terminations that allow for higher current redistribution will be investigated in future work.

8.2.3 Current source and quench detection

The current source is a 12 pulse thyristor converter from ABB which was originally designed as source for large superconducting fusion magnets. It provides a maximum current of 50 kA DC and up to 30 V.

The current of the power source was measured with a Fiber Optic Current Sensor from ABB which is based on the Faraday effect. It has a rated current of 80 kA, a sampling rate of 4 kHz, an accuracy of ±0.1 %, and a linearity of ±0.1 %. The Fiber Optic Current Sensor outputs a ±1 V signal which is recorded in the data acquisition.

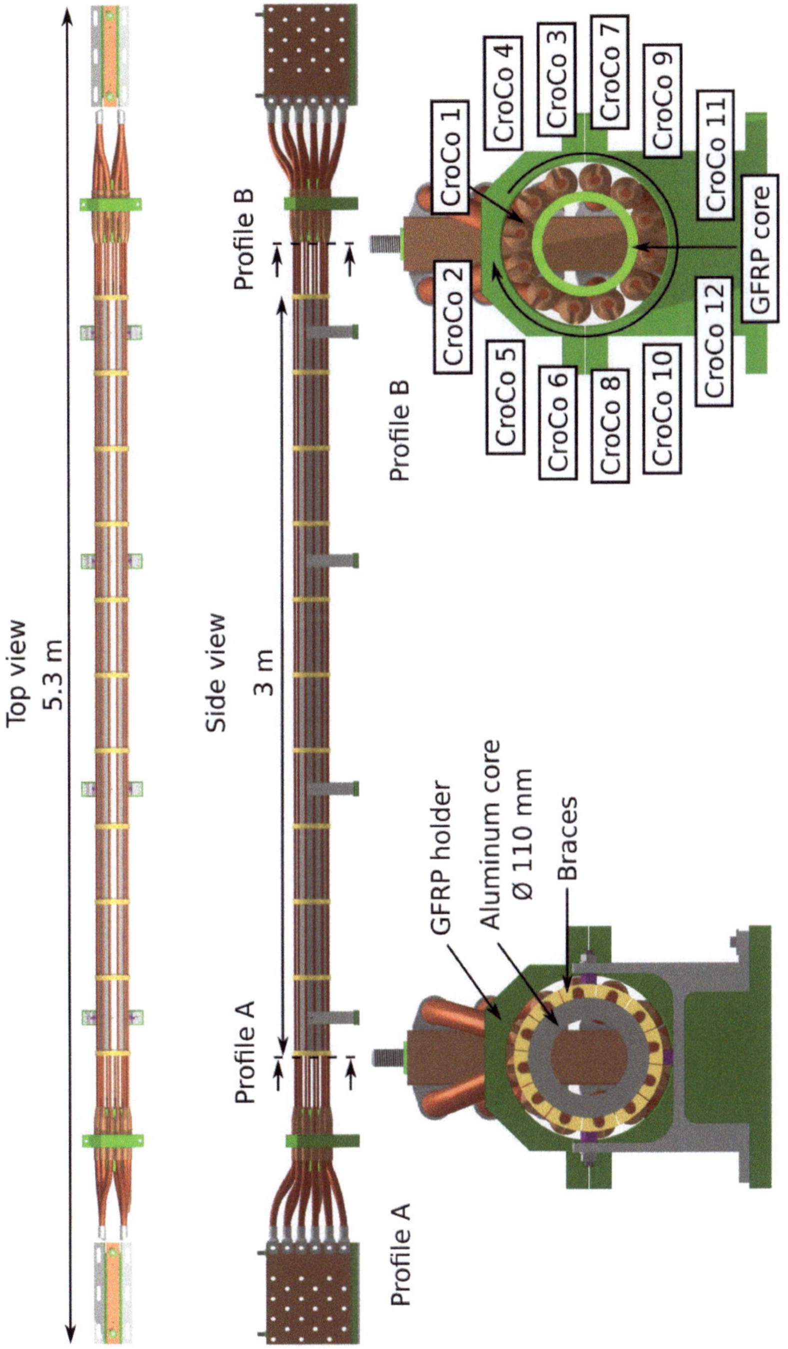

Figure 8.14: Design schematic of the 35 kA cable demonstrator cable

The quench detection is realized by a RC low pass filter coupled with a threshold switch. Thresholds can be varied between 1 mV and 10 mV in 1 mV steps. The quench detection triggers the emergency shut down of the power source which initiates a ramp down with a rate of 1 kA/ms. Details on the quench detection are published in [Hol+13].

8.2.4 Data acquisition

The experiment requires a series of voltage measurements with an expected minimum voltage drop between 200 µV and 300 µV across the CroCos at their critical current. Therefore, a measurement system with a sensitivity in the range of 1 µV is necessary. A differential measurement method was used due to the small signal and the relatively long cables. All measured voltages are floating signals for which isolated measurement channels are beneficial. Based on previous experiments, significant current ripples are expected from the current source. To limit the impact of these ripples on the measurement a low-pass filter of less than 50 Hz was deemed necessary since the converter was not designed with extended current smoothing capabilities.

The NI SCXI 1125 conditioning module in combination with the NI PCI 6280 analog-digital converter offers the aforementioned requirements and was chosen as measurements system in this project. The SCXI 1125 multiplexes several channels into one differential measurement channel which is read by the NI PCI 6280.

The conditioning module is equipped with a 4 Hz or 10 kHz low pass filter. For the majority of the measurements the 4 Hz filter was active. A sampling rate of at least 8 Hz would be necessary to fulfill the Nyquist-Shannon-Criterion. The effective sampling rate was set to 100 Hz to avoid in any concerns in regards to aliasing,

The NI PCI 6280 is a 18 bit measurement system with a variable input range between ±10 V and ±0.1 V resulting in a sensitivity of 0.8 µV at an input range of ±0.1 V.

8.3 Demonstrator cable test

8.3.1 Cryostat cool down and warm up

Figure 8.15 a) shows the cryostat wall temperatures over time leading up to an experiment and the warm up phase afterwards. The different temperatures reflect the position of the temperature sensors in the cryostat as shown in Figure 8.12.

During the cool down the temperature difference across the temperature sensors is kept roughly within 100 K to limit thermal strain on the cryostat. Temperature oscillation (Figure Figure 8.15 b) can be observed during the first 4 h to 6 h of the cool down cycle in particular at the bottom temperature sensors. The oscillation is caused by gas bubbles in the LN_2 supply line that temporarily reduce the LN_2 flow rate. The amount of gas bubbles reduces over time a)

b)

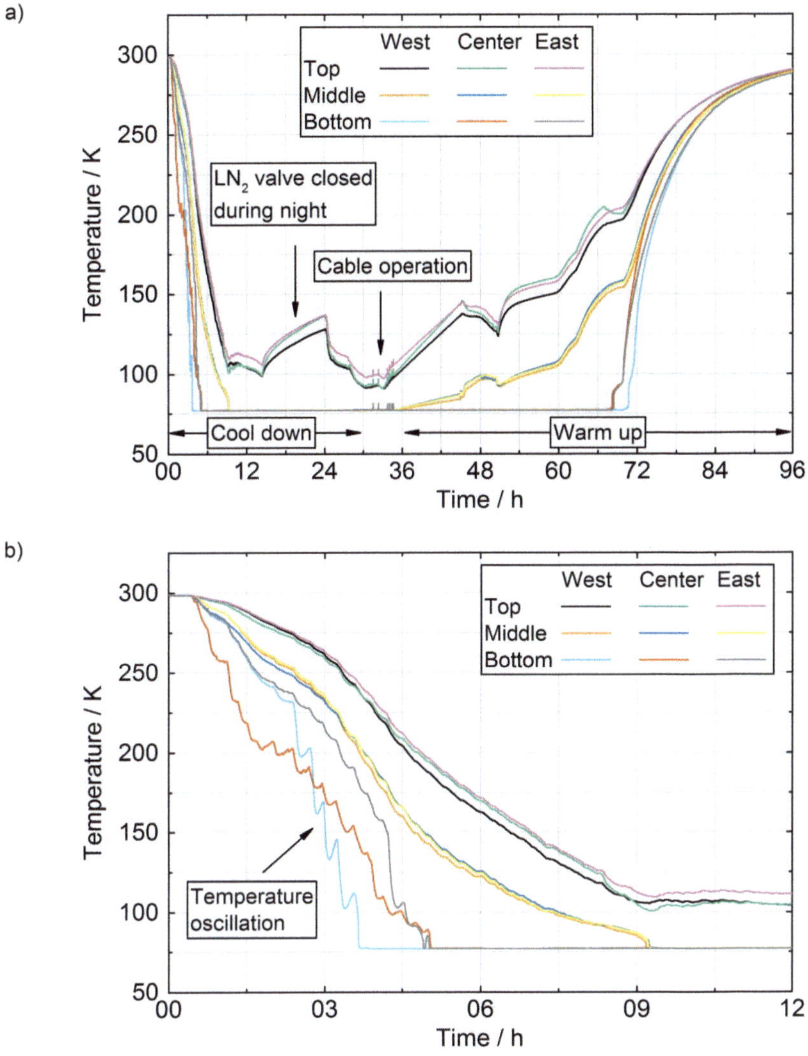

Figure 8.15: a) Temperature development within the cryostat during an experimental cycle b) Detail of Figure 8.15 a for the first 12 h of the cooling cycle with emphasis on the temperature oscillations

as the supply line itself cools down. Temperature sensors on the *West* side tend to show lower temperatures over time which indicates a small tilt of the cryostat.

The cable operation is distinguishable by simultaneous peaks in all temperature sensors. The peaks coincide with the triggering of the quench detection and the subsequent emergency shutdown and are therefore likely an induction based measurement error and do not represent an actual temperature rise.

The warm up takes about 72 h if no external heaters are used and at least another 24 h to remove condensed water from the cryostat and the sample. The warm up phase is initiated by closing the LN_2 supply line and pumping room temperature gaseous nitrogen into cryostat in order to reduce the accumulation of water within the cryostat. The warm-up phase could likely be significantly shortened by the use of dedicated heaters but no such attempts have been made so far.

The total losses of the cryostat were estimated by measuring the LN_2 level in the cryostat before the LN_2 supply was stopped for the night and again in the morning before the supply was resumed. The average losses during this 12 h stand-by period amounted to about 10 g/s LN_2 which is about half of the estimated value. The large deviation is due to worst case assumptions in the loss calculation that were adopted as a safety measure in the design of the water bath heaters.

8.3.2 Cable test

The current source was operated in such a way that the individual steps of the measurement programs were entered manually. The current is ramped up to a constant value where it rests for at least 30 s until the current is ramped to the next constant value. This proceeds until the previous defined quench detection limit is reached or the current is manually ramped down again.

Table 8.3 lists the maximum constant current that was achieved in each test run. During the first two measurements the quench detection triggered after 25 kA and 23 kA respectively. It is important to note that voltage taps used for the quench detection include the terminal resistance of the CroCos as shown in Figure 8.2. A total terminal resistance of 0.5 μΩ (Tab. 8.2) at a current of 3.5 kA (Tab. 8.1) would already result in a voltage drop of 1.75 mV. The voltage drop over the remaining CroCo length (3.4 m) at its critical point (100 μV/m) would be 0.34 mV. The quench limit was therefore initially set to 3 mV in order to measure overcritical behavior as well.

Table 8.3: High current experiments

Measurement	Maximum current	QD integration time	Maximum QD limit
1	25 kA	10 ms	3 mV
2	23 kA	10 ms	3 mV
3	33 kA	50 ms	3 mV
4	34 kA	50 ms	4 mV
5	34 kA	50 ms	5 mV
6	35 kA	50 ms	6 mV

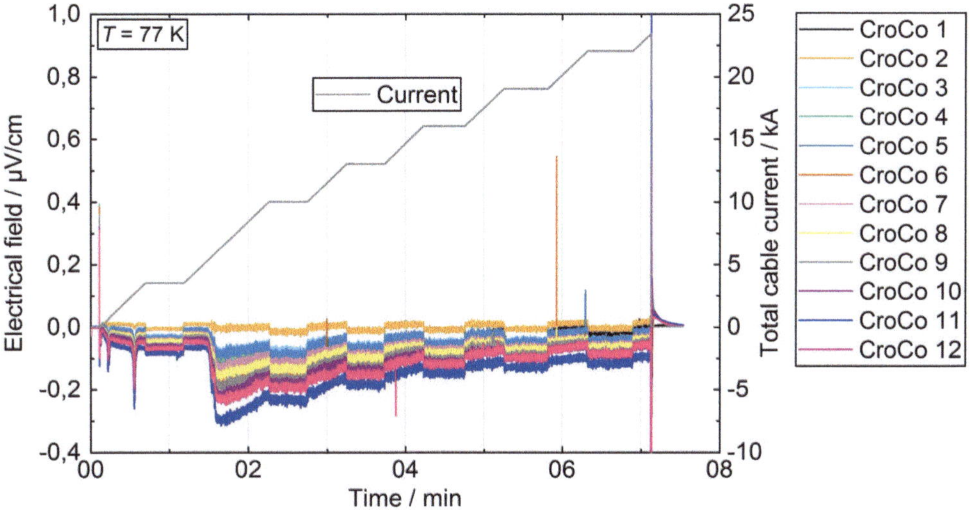

Figure 8.16: Electric field across all CroCos as a function of time and at various total cable currents during the second measurement run

In Figure 8.16 the electrical field of each CroCo over the 2 m voltage taps is displayed as a function of time at various total cable currents. At the beginning, several small peaks across all CroCos are visible as the power supply is switched. At around 1:30 min a drop in the electric field of most CroCos is observable. This drop is associated with a change in the rectifying program of the power supply at about 7 kA. The exact change in behavior is unknown as it is proprietary information of the power supply manufacturer. If one compares the magnitude of the change with the position of the CroCos around the cable core (see Figure 8.14) than it is noticeable that the CroCos with the largest change in the electrical field are all located at the bottom. A possible causal connection needs to be further investigated in future measurements.

The voltage drop caused by the change in switching behavior recovers to some degree as the current increases. At 23 kA the emergency shut down is triggered by the quench detection even though none of the CroCos show any indication of quenching. The likely cause for the emergency shut down are the random peaks that already occurred at lower currents in CroCo 5, CroCo 6 and CroCo 12.

The data displayed in Figure 8.16 is already averaged as described in Section 8.2.4 meaning that the measured peak voltage is likely significantly higher and thus triggering the quench detection which is based on a raw signal. The integration time of the QD was subsequently increased from 10 ms to 50 ms in order to avoid false-positive emergency shut downs. The increased integration time of the quench detection made it possible to achieve currents of up to 33 kA. In order to achieve higher currents its was necessary to incrementally increase the quench detection limit of CroCo 2 and CroCo 12.

Figure 8.17 shows the electric field over the 2.8 m voltage taps of all CroCos during a measurement with a maximum total cable current of 35 kA. Towards higher cable currents the initial drop in the electric field at around 7 kA total cable current recovers to the expected zero until the CroCos reach the flux flow region.

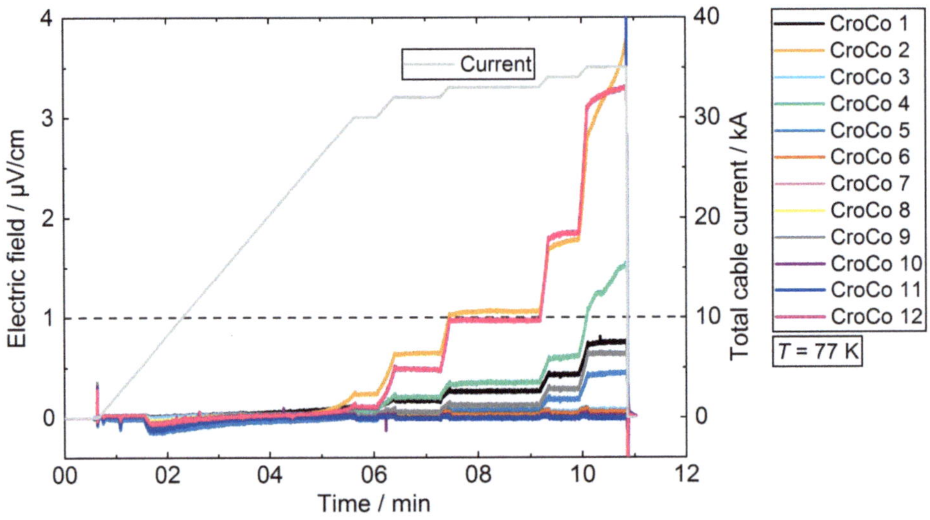

Figure 8.17: Electric field across all CroCos as a function of time and at various total cable currents during the sixth measurement run

The first CroCos that develop a distinctive electric field are CroCo 2 and CroCo 12. At 33 kA both of these CroCos reach their critical current but can still achieve a steady state.

At 35 kA total cable current CroCo 12 does not achieve steady state and develops a run away behavior. The experiment was manually shut down after about 1 min at 35 kA total cable current. Also, CroCos 1, 4, 5 and 9 entered the flux flow region at 35 kA and display a significant electric field. In particular CroCo 4 is interesting as it does not achieve steady state at 35 kA even though the electric field of CroCo is about 2 µV/cm lower compared to CroCo 2 or CroCo 12. During single CroCo characterization CroCo 4 achieved a steady state at 3.5 kA with an electrical field of around 2.5 µV/cm.

Figure 8.18 shows the individual current in each CroCo as a function of the total cable current up to 35 kA and in dashed lines the expected behavior as shown in Figure 8.10. In general, the current distribution follows the expected linear distribution and only shows non-linearity for CroCo 2 and CroCo 12 for currents beyond 30 kA.The small nicks in the measured curves are due to inductive elements that disappear when the measurement enters a constant current section.

There are several deviations compared to the simulation performed in section 8.1.3. The spread in the current distribution is higher than expected. This is immediately recognizable for CroCo 11, that carries significant less current than all other CroCos.

The simulation correctly identified CroCo 2 and CroCo 12 as current limiting for the entire cable. However it predicted CroCo 12 to be the critical one and not CroCo 2 which in turn can be attributed to small deviations in the contact resistance.

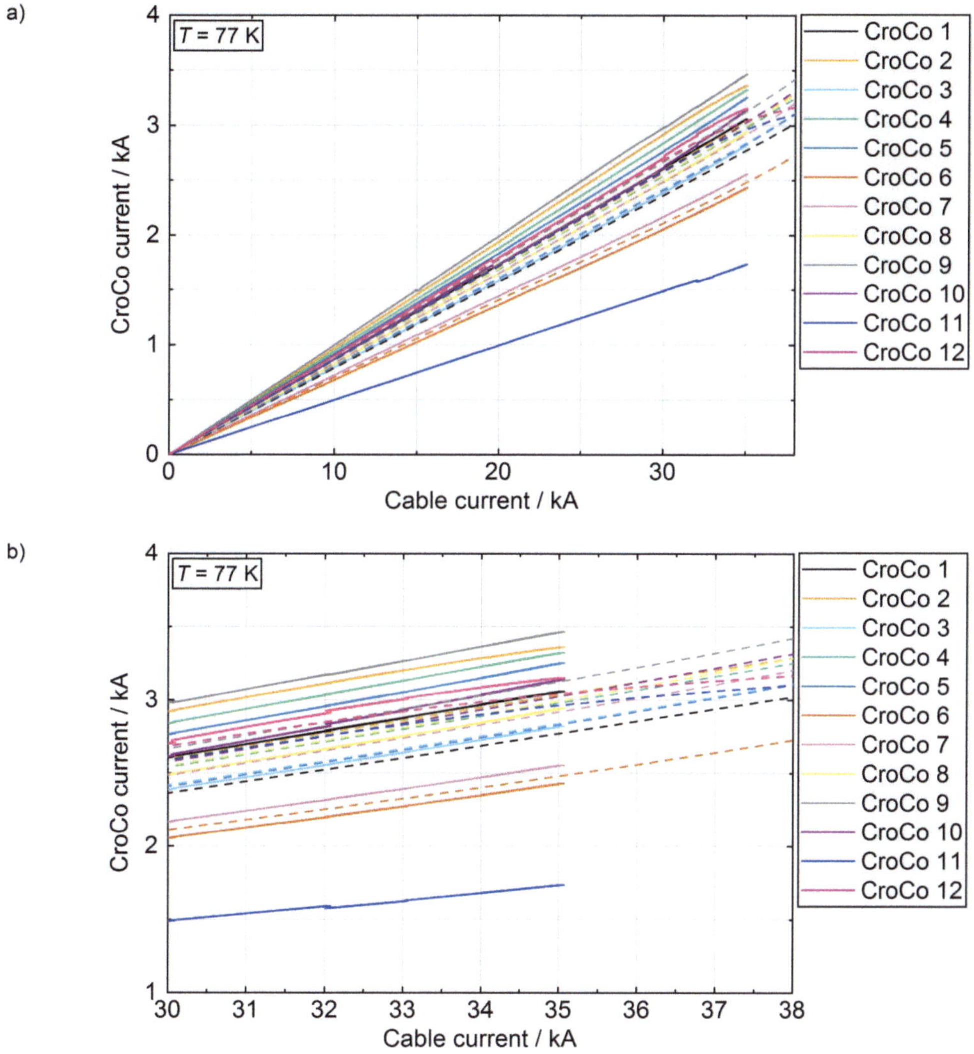

Figure 8.18: a) Individual CroCo current as a function of the total cable of up to 35 kA b) Detail of Figure 8.18 a above 30 kA. Compact lines represent measured values during the 35 kA measurement. Dashed lines show the expected values based on measured contact resistances in the single CroCo characterization (see Table 8.2)

For the calculations shown in Figure 8.10 it was assumed that the contact resistance R_{end} between the termination's cable shoe and the copper block (see Figure 8.14) are identical for all CroCos with a value of 0.5 $\mu\Omega$. Based on the measured current distribution one can calculate the contact resistance between the cables shoes and the copper block by incrementally increasing the contact resistance of the CroCos until the measured current distribution is achieved. Figure 8.19 shows a well matching current distribution simulation (dashed lines) compared to the measurement results (compact lines). This confirms that current distribution can be simulated very well with the data of the individual CroCos.

Table 8.4 summarizes the numerically calculated contact resistance during the experiment shown in Figure 8.17. The lowest contact resistance was assumed to be 0.5 $\mu\Omega$ based on prior measurements. In general, a spread of contact resistances can be observed with the highest (3.18 $\mu\Omega$) being an order of magnitude higher than the lowest calculated value (0.64 $\mu\Omega$) with an additional outlier at 9.06 $\mu\Omega$.

The high resistance at CroCo 11 was likely caused by water condensing between the cable shoe and copper block within the cryostat during prior thermal cycles tests and thus reducing the effective cross section available for current transfer. The same issue was observed to a lower degree in all connections. A possible solution could be in polishing the cable shoes and copper block surface prior to the experiments.

Table 8.4: Calculated contact resistances based on measurements during cable operation at 77 K

CroCo	1	2	3	4	5	6	7	8	9	10	11	12
R_{end} / $\mu\Omega$	1	0.64	2.05	0.89	0.88	0.5	3.18	2.72	0.82	1.96	9.06	1.86

8.3.3 CroCo performance post cable operation

One major concern in the development of a cable prototype is the degradation of the superconductor due to for example inhomogeneous cool down resulting in damaging strain. Therefore, it is prudent to investigate if degradation occurred during or after the experiments performed in this work. For this purpose, CroCo 2 was dismounted from the cable assembly and measured again at 77 K with the setup described in Section 8.1.1. CroCo 2 was chosen as it was one of the CroCo that quenched during the 35 kA experiment (Figure 8.17) and was easily accessible (Figure 8.14).

Figure 8.20 shows the electrical field of CroCo 2 as a function of the current at various experiments. The black line represents the result of the initial CroCo characterization already shown in Figure 8.3. The brown line shows the electrical field of CroCo 2 during the 35 kA experiment shown in Figure 8.17 which exactly matches the initial CroCo characterization. In blue is shown a measurement after the cable measurements were performed and CroCo 2

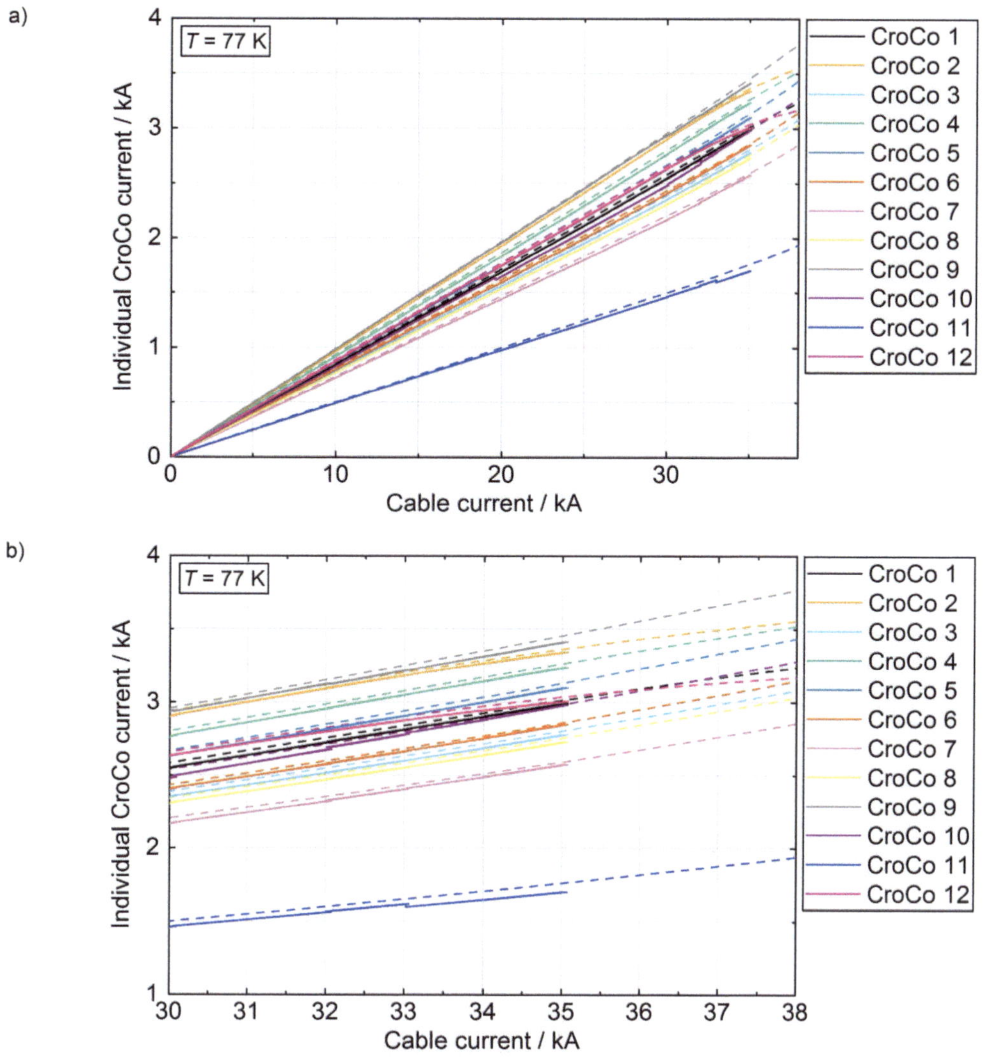

Figure 8.19: a) Individual CroCo current as a function of the cable current up to 35 kA b) Shows a detail of Figure 8.19 a) above 30 kA. Compact lines represent the measured values. Dashed lines show the simulated values with incrementally increased contact resistance (see Table 8.4).

disassembled from the cable demonstrator. The CroCo shows an increased critical current by about 32 A which is about 1 % of the total current and is therefore assumed to be within measurement uncertainty.

Table 8.20 summarizes the critical currents of CroCo 2 during the above mentioned experiments. The variation in n value between the CroCo 2 Single and CroCo 2 Cable appear to be

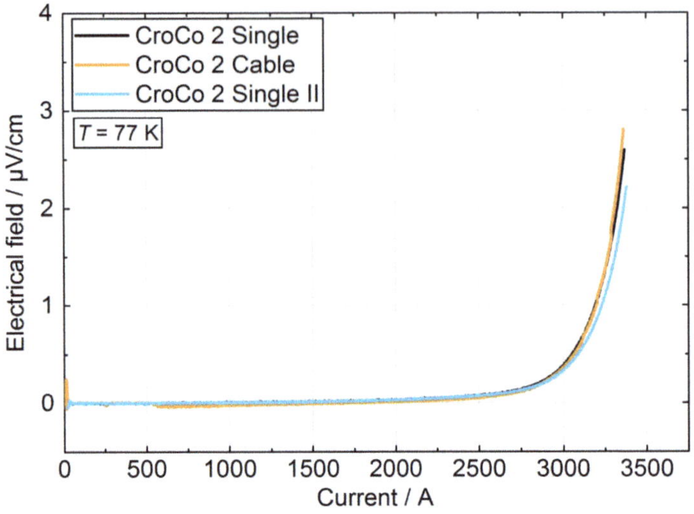

Figure 8.20: Electrical field as a function of current of CroCo 2 during initial CroCo characterization, within the cable assembly, and in a single CroCo arrangement after the cable experiments

largely caused by fitting algorithms as no major deviation between the curves can be recognized in Figure 8.20.

Table 8.5: Summary of electrical properties of CroCo 2 during single characterization and cable experiments

Measurement	I_c / A	n
CroCo 2 Single	3193	16.4
CroCo 2 Cable	3189	18.5
CroCo 2 Single II	3225	15.8

In Table 8.6 the contact resistances are shown prior and after the cable experiment. The contact resistances of the superconductors remain unchanged while some variation is observed at the cable shoe of the minus pol and between the connection of copper litz and copper cylinder. The variation might be cause by movement of the copper braces that are used as voltage tap (Figure 8.2) during the assembly and disassembly of the cable.

In summary, no indication for degradation was found in CroCo 2 which experienced multiple cooling cycles and quenched multiple times during cable operation.

Table 8.6: Summary of contact resistances of CroCo 2 before and after the cable experiments

CroCo	CL+	CL-	CS-CL+	CS-CL-	CL-CC+	CL-CC-	CC-CroCo+	CC-CroCo-
-					\ μΩ			
Single	2.93	2.71	0.64	0.62	0.47	0.4	0.18	0.16
Single II	3.04	2.72	0.64	0.45	0.45	0.56	0.18	0.17

8.4 Chapter summary and outlook

In this chapter the individual characterization results of 12 3.6 m CroCos long was shown. The critical current of each CroCo was measured with two voltage taps with a distance of 2.8 m and 2 m, respectively. The measured critical currents were then compared with critical current calculations that showed a good agreement when the average critical current of the *RE*BCO tapes is used as an initial point for the iterative calculation.

Microscopic cross sections of each CroCo showed severe deformations in the first CroCo manufactured but a steady improvement in regards to the roundness of the CroCo and shape of the cross formed by the *RE*BCO tapes for the following CroCos. The rupture of one *RE*BCO tape resulted in no lasting consequences in regards to the shape of the CroCo.

The resistances of the CroCo terminations were measured in detail. The resistances between CroCo and copper cylinder proved very low resistive with values between 0.15 $\mu\Omega$ and 0.32 $\mu\Omega$. Based on the termination resistances and the critical currents of the CroCos it was possible to calculate an expected current distribution and maximum current for the cable demonstrator that consists of all 12 CroCos connected in parallel.

The cable demonstrator was successfully assembled, cooled down and a series of experiments were performed. The final experiment achieved a steady state at 34 kA and for a short period 35 kA. The current distribution did not match the expected current distribution. The cause was likely condensed water between contacts.

In order to check for degradation one of the CroCos that quenched during the demonstrator cable experiment was removed from the demonstrator and measured again individually. No degradation was found.

The current layout of the demonstrator cable allows for very little current redistribution when one of the CroCos approaches its critical current due to the high termination resistances. In future work it is planned to decrease the termination resistance to the very low resistive connections between the CroCo and copper cylinder thus allow for current redistribution. This measure could increase the current carrying capacity of the demonstrator to more than 40 kA. In another experiment it is planned to reduce the diameter of the aluminum core and therefore decrease the size of demonstrator to a more compact design.

9 Application of HTS DC cables in aluminum plants

In the following the application of HTS high current DC cables is discussed at the example of an aluminum smelter. The focus of this design study is the superconducting cable in general and the superconductor itself in particular due to the fact that the *RE*BCO tapes are expected to be the major cost factor in the superconducting cable system.

9.1 Aluminum production

The production of primary aluminum is based on an electrolysis process (Hall-Héroult process) in which aluminum oxide (Al_2O_3) is dissolved in an electrolyte consisting of molten cryolite (Na_3AlF_6) and aluminum fluoride (AlF_3). The oxygen of aluminum oxide reacts at the carbon anodes to carbon monoxide in a first step and then further to carbon dioxide thus consuming the anodes eventually. [KD14]

The overall reaction formula is

$$Al_2O_3 + 3\,C \longrightarrow 2\,Al + 3\,CO \tag{9.1}$$

Within an aluminum plant several hundred electrolysis cells are arranged side by side and connected in series which results in facilities that can be 1 km long and more. The voltage in the electrolysis cells may vary slightly depending on the heat balance, or cell operating conditions etc. The current however is constant across all cells and ranges between 100 kA and 600 kA. [Tri18; KD14]

The electrolysis cells are operated non-stop all year around since an interruption could lead to the freezing of the electrolyte which would entail a time consuming and therefore costly restarting procedure [RUN95].

The energy consumption in the electrolysis process is about 13 kWh/kg in modern aluminum plants. For example, TRIMET Aluminum SE operates three aluminum electrolysis plants in Germany and produced 385000 t of aluminum in 2018 [Tri19]. This equals an electricity consumption of about 5 TWh in 2018. As a perspective, about 543 TWh of electricity were produced in Germany in 2018 [Bur19] which means that about 1 % of the total electricity production in Germany was consumed in three aluminum plants. Naturally, the electricity price in such energy intensive industries is one of major cost factors. In the following, it will be investigated if superconducting high current cables can contribute to energy and cost savings.

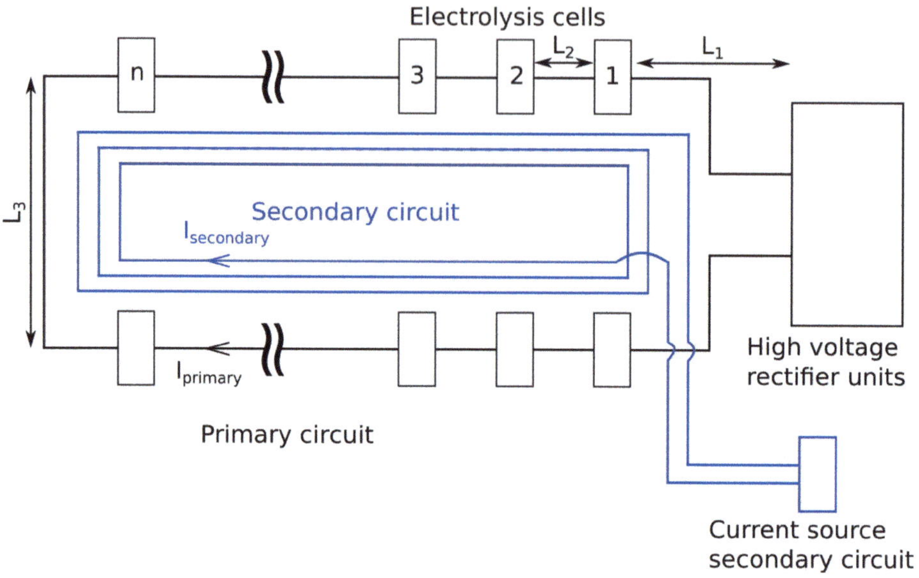

Figure 9.1: Schematic of primary and secondary circuit within an aluminum production plant (after [Duv+14, Fig. 7])

9.2 Superconducting cable systems within aluminum plants

9.2.1 Superconducting cable use cases

Energy savings with the help of superconductors in the context of aluminum production is often associated with substituting conventional aluminum or copper bus bar systems with superconducting ones [RUN95; Mor15]. Energy savings are possible due to the extremely high currents (up to 600 kA) coupled with nearly non-stop operation of the electrolysis cells.

Previously two different circuits within an aluminum production plant have been examined for the use of superconducting high current cables [RUN95; Mor15]. These are illustrated in Figure 9.1 with the first one being in the primary circuit that connects rectifiers units with electrolysis cells. Within the primary circuit, the current can reach the above mentioned 600 kA. However, one has to acknowledge that a superconducting cable is a more complex system than a conventional air cooled bus bar and the risk of failure is significantly higher. A back-up system would be necessary due to high system failure cost. [RUN95]

Potential candidates for replacements within the primary circuit are aluminum bus bars between the rectifiers and the electrolysis cell which can be up to several hundred meters long (L_1 in Fig. 9.1). The connections between the aluminum cells in the typical U-shape arrangement

L_3 can also be several dozens of meter long. The connections between the electrolysis cells L_2 should not exceed a couple of meters and therefore an unlikely use case. [Mor15]

A secondary circuit can be placed within the aluminum smelter to mitigate the magnetic field generated by the primary circuit. This can be beneficial as the magnetic field can have an adverse effect on the performance of the electrolysis cell [Ren+16]. The current in the secondary circuit can vary between 10 % and 150 % of the electrolysis current depending on the design [RUN95; Ren+16]. The operational requirements are not as high in the secondary circuit as in the primary circuit and therefore no additional back-up system is necessary [RUN95].

The secondary circuit is a particular interesting use case for a high current cable consisting of CroCos as the CroCos do not necessarily need to be connected in parallel. Instead the CroCos can be placed within one cryostat and connected in series creating multiple turns as it was proposed in [Duv+14; Mor15]. This way the same magnetic field is generated at the electrolysis cell but the losses at the terminations can be drastically reduced due to the lower current that is transferred to LN_2 temperatures (see Section 9.6.2).

9.2.2 Cryogenic system

A superconducting cable needs a cryogenic system that is capable of removing any heat that is generated or leaked into the cable. In the following a brief summary of the available concepts is given.

The cable discussed in this case study is based around a forced flow cooling concept meaning that a coolant (LN_2) is pumped along the cable to remove heat that leaked into the system. In cable systems the nitrogen usually flows in a closed system meaning that it does not evaporate into the environment under operating conditions. Instead the coolant is continuously re-cooled by a cryogenic system thus maintaining operating conditions of the cable.

These re-cooling systems can be grouped in closed and open systems. Closed systems are self contained units that only require electricity, maintenance and often cooling water. Open systems are based on the evaporation of LN_2 and therefore need a continuous supply of LN_2 in addition to electricity and maintenance.

Closed systems

One method of re-cooling the circulated LN_2 is by using commercial cryocoolers and coupling them through for example a heat exchanger with the circulating coolant. The cryocoolers will manage the temperature within the heat exchanger and remove any excess heat from the system.

Table 9.1 lists some basic properties of currently commercially available cryocoolers from several manufacturers. As can be seen, cryocoolers based on a Stirling process are readily available for several kilo watts of cold power. One significant drawback of the Stirling cryocoolers, is the relatively short maintenance interval. Cryocoolers based on a Gifford-McMahon process perform better in terms of maintenance but provide also significant less cold power. The pulse tube cryocooler are almost maintenance free due to avoiding any moving parts in the cold system but the cold power of currently available pulse tube cryocoolers is limited. This is however not an inherent problem of the technology as was shown in a demonstrator pulse tube cryocooler that provided 1 kW at 77 K for a superconducting cable [Che+10]. Even larger cryocoolers with 10 kW capacity based on a reverse turbo Brayton process are currently under development [Cha+].

Table 9.1: Basic performance overview of commercially available cryocoolers

Cryocooler	Cold power at 70 K	Input power	Maintenance interval	Process	Reference
SPC-4	3200 W	46 kW	6000 h	Stirling	[Sti19]
AL600	500 W	12.5 kW	10000 h	Gifford-McMahon	[Cry19]
CH-110	160 W	N/A	13000 h	Gifford-McMahon	[SHI19]
PT90	70 W	4.3 kW	20000 h	Pulse Tube	[Cry19]

Open system

A different approach is to utilize LN_2 to re-cool the circulating coolant. The concept is to create and maintain a vacuum within a LN_2 tank (sub-cooler). The temperature of the LN_2 within the tank will decrease as the pressure decreases since the nitrogen boiling temperature is pressure dependent (see Figure 6.1). The sub-cooler can then be coupled with the circulating coolant to maintain the operating temperature.

The cold power Q provided by such a sub-cooler is dependent on the mass flow $\dot{m}$ of the evaporating nitrogen which can be calculated by

$$\dot{m} = \frac{Q}{\Delta h_{vap} - \Delta h} \tag{9.2}$$

with Δh_{vap} being the enthalpy of vaporization of nitrogen of about 200 kJ/kg and Δh the enthalpy change of the LN_2 due to subcooling. For example a sub-cooler providing 4000 W at 70 K would require

$$\dot{m} = \frac{3200\,\mathrm{W}}{200\,\mathrm{kJ/kg} - (136.97 - 122.75)\ \mathrm{kJ/kg}} = 62\,\mathrm{kg/h} \tag{9.3}$$

assuming the liquid nitrogen is supplied at 77 K. To liquefy nitrogen consumes about 0.48 kWh/kg of electricity in large scale system resulting in about 30 kW electricity consumption for the above mentioned 4 kW at 70 K [Her+16].

The power consumption of the pumps maintaining the vacuum within the sub cooler also needs to be considered. The mass flow of 62 kg/h liquid nitrogen is equal to about 78 m^3/h of gaseous nitrogen. This kind of suction capacity could be realized for example with a rotary claw vacuum pump that have an electricity requirement of about 4 kW.

The maintenance interval of the sub-cooler system is based on the maintenance requirements of the vacuum pumps which require in this example an oil change every 20000 h and a general overhaul every 6 years.

Comparison

Table 9.2 lists a basic comparison of the two discussed re-cooling systems. The overall losses within an open system are usually lower than the closed system due to the increased efficiency of large scale liquefier units. If the lower losses also translates into lower operational cost depends on the electricity price.

In terms of space and weight the closed system have a clear advantage due to avoiding the large LN_2 tanks that are necessary for the open system.

Both systems can be designed in a redundant manner [Ste15]. In a closed system the entire system needs to be redundant while it is sufficient to install multiple vacuum pumps in the open system to create redundancy [Her18]. Which case is more cost efficient is of course dependent on the cold power requirements etc.

The maintenance intervals of the closed system depends on the chosen cryocooler process which in turn largely depends on the cold power demand. The maintenance of the vacuum pumps also depends on the cold power demand since the vacuum pump design is a function of the mass flow through it.

Table 9.2: General comparison between the open and closed cryogenic re-cooling systems

Parameter	Closed system	Open system
Overall losses	-	+
Space & weight	+	-
Redundancy	+	+
Maintenance	-/+	+

One additional issue for the open system can be the supply of LN_2 if the aluminum plant is in a remote location which is often transported to the facility via trucks. On site, the liquid nitrogen can be stored in tanks that are available with capacities of 60000 kg [Lin19].

9.2.3 Current leads

One of the major heat sources in relatively short but high current superconducting cable systems are current leads. Current leads transfer electrical current from ambient to cryogenic temperatures while minimizing the heat load into the cold system. A thorough analysis of various current lead concepts in regards to their cold losses and their ideal power consumption at ambient temperature was performed in [Gol04]. Table 9.3 is assembled and translated into English from various tables contained within [Gol04].

Table 9.3 provides an overview of the performance of several different current lead concepts. Listed are heat loss at cryogenic temperatures per current flowing through the current lead and minimum amount of power necessary to provide the cold power based on the ideal Carnot process (see Section 9.6.2) with an ambient temperature of 293 K.

Table 9.3: Loss overview of different types of current leads assuming an ambient temperature of 293 K. A copper *RRR* of 100 was used for the conduction based current leads. The gas cooled current assumed a heat transfer coefficient of 1000 W/(m² K) (assembled and translated into English from [Gol04])

Type	Cold losses W/kA	Ideal power input W/kA
Conduction, One-Stage	44.23 at 77.4 K	123.2
Conduction, Two-Stage	31.86 at 154 K 22.13 at 77.4 K	Σ90.1
Conduction, Three stage	24.72 at 197 K 17.17 at 150 K 17.44 at 77.4 K	Σ83.9
Conduction, Infinite stages	10.4 at 77.4 K	Σ78.4
Vapor cooled	27.2 at 77.4 K	Σ101.7

The simplest version of a current lead is a one stage conduction based current lead in which a resistive conductor is connected to the superconducting cable. In this case heat will propagate via conduction into the superconducting system where it is removed by the cryogenic system. In [McF59] an optimization process for minimizing the heat leakage into the cold system of these so called conduction based current leads is described. All conduction based current leads listed in Table 9.3 utilize copper with a residual resistance ratio (RRR) of 100. However copper is not mandatory since [McF59] showed that the heat leak of an optimized current lead is largely independent of the material properties as long as the material follows the Wiedemann-Franz-Relation.

A method of reducing heat leakage in conduction based current leads is to introduce additional temperature stages between the operating temperature and the ambient temperature. The benefit is based on the fact that it is less energy intensive to remove heat at these intermediate temperatures. Table 9.3 lists a two-, three-, and infinite stage current leads. The temperature levels of the stages are chosen based on an optimization process that minimizes the heat leakage into the system [Gol04]. Introducing a second stage already reduces ideal power consumption by roughly a third. However it can also be seen that adding further temperature stages quickly develops diminishing returns with only a decrease in ideal power consumption of 7 % for a third temperature stage.

The ideal current lead would consist of an infinite amount of temperature stages and is listed in Table 9.3.

If the temperature at the lower end of a current lead is maintained by for example by a LN_2 bath then the evaporated gaseous nitrogen could be used to cool the current leads and thus reducing the losses. This so called vapor cooled current lead is the last entry in Table 9.3. The listed current lead refers to a self-sufficient vapor current lead in which no additional gaseous nitrogen is added to increase the cooling capacity. From Table 9.3 it can be seen that a self-sufficient vapor cooled current lead has about 20 % less energy consumption than a one stage pure conduction cooled. One challenging aspect of vapor cooled current leads is to ensure a sufficient heat transfer coefficient which is assumed to be 1000 W/(m² K) in the listed scenario.

9.3 Primary circuit

Likely candidates for a superconducting cable in the primary circuits are the connections from the rectifier units to the electrolysis cells (L_1 in Figure 9.1) and between pot lines (L_3 in Figure 9.1).

Design specifications

For the purpose of this design study a superconducting cable with an operational current of 100 kA is considered for the primary circuit. In Section 8.3 Figure 8.17 it was shown that is possible to operate CroCos at their critical current and even beyond it, provided there is sufficient cooling. Operating a 100-kA class cable at its critical current however would generate losses equal to

$$q_{SC} = E_c \cdot I = 100\,\mu\text{V/m} \cdot 100\,\text{kA} = 10\,\text{W/m} \tag{9.4}$$

which would be roughly 10 times higher than the cryostat losses. The losses directly transfer into operating cost for the superconducting cable and therefore the exact operating point of a superconducting cable needs to be determined in an economical evaluation. For the purpose of this design study the losses will be defined as q_{SC} = 0.1 W/m which amounts to roughly 10 % of the cryostat losses. The operational to critical current ratio is then equal to

$$\frac{I_{op}}{I_c} = \left(\frac{q_{SC}}{E_c I_{op}}\right)^{1/n} = \left(\frac{0.1\,\mathrm{W/m}}{100 \cdot 10^{-6}\,\mathrm{V/m} \cdot 100000\,\mathrm{A}}\right)^{1/20} = 0.8 \tag{9.5}$$

Therefore the critical current of the cable system needs to be at least 125 kA.

The operational voltage of 2 kV is based on the maximum voltages commonly encountered in the primary circuit [FI02].

The inlet temperature of the LN_2 is limited by the nitrogen triple point and the fact that a temperature gradient between the re-cooler and the inlet temperature is necessary to maintain the temperature. Considering the inlet temperature of previous superconducting cables (Table 5.7) shows that an inlet temperature of 67 K and a temperature rise to 70 K appears to be within reasonable bounds.

The maximum permissible temperature rise during a short circuit is largely dependent on the pressure of the coolant and therefore a high inlet pressure and low pressure drop are desirable. The maximum pressure of LN_2 is limited by its critical point, the approved maximum pressure of the cryostat (2 MPa for Nexans Cryoflex [Nex]) and the circulating pumps (1.5 MPa [Zer12]). An inlet pressure of 1.5 MPa is chosen for this design study. The pressure loss is set to 0.001 MPa due to the large cross section available for the coolant channel.

The short circuit specification for the primary circuit are chosen to be rather restrictive due to the large currents flowing. Sophisticated current limiter are likely necessary to manage the defined fault duration of 0.1 s

The inner diameter of the cable is limited to 130 mm to ensure that the 500 m cable can be transported on a single drum without any major restrictions on transportation.

Table 9.4 summarizes the design specification and boundary conditions.

Table 9.4: Design specifications and boundary conditions used in this case study

Description	Symbol	Value	Description	Symbol	
Operating current	I_{op}	100 kA	Inlet temperature	T_{in}	67 K
Operational voltage	U_{op}	<50 V	Outlet temperature	T_{out}	70 K
Cable length	L	500 m	Inlet pressure	p_{in}	1.5 MPa
Fault current	I_{fault}	$5 \cdot I_{op}$	Outlet pressure	p_{out}	1.499 MPa
Fault duration	t_{fault}	0.1 s	Operational to critical current ratio	$I_{\mathrm{op}}/I_{\mathrm{c}}$	0.8
Ambient temperature	T_{H}	293 K			
Inner cryostat diameter	d_{cryo}	130 mm			

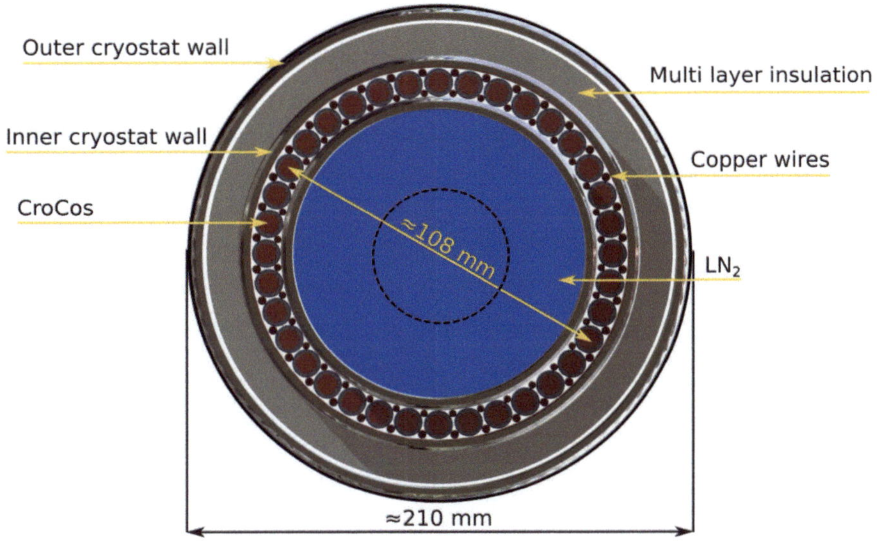

Figure 9.2: Superconducting primary circuit cable

Conceptual design

Following the design process outlined in Chapter 6 yields a cable as described by its characteristics in Table 9.6 and schematically shown in Figure 9.2. The first step in the conceptual design process is the calculation of the dielectric thickness. In this case, the relatively low voltage of 2 kV is insulated by a layer of 1 mm thick PPLP between the CroCos and inner cryostat wall.

Next in the design process is the calculation of the stabilizer requirement. The maximum permissible temperature rise during a fault is determined by the boiling temperature of the liquid nitrogen at a pressure of p_{out} = 1.499 MPa which is about 110 K (see Figure 6.1). Adding a safety margin of 2 K results in T_{end} = 108 K. Under the given boundary conditions a stabilizer cross section of 960 mm^2 is necessary (using the code in Annex A.3) to limit the temperature rise during fault.

In section 6.8 Table 6.6 it was shown that the usage of CroCos consisting of 6 mm and 4 mm wide tapes is beneficial under the boundary conditions imposed in by the design process. Accordingly the cable utilizes in total 36 CroCos 6/4 CroCos in parallel connection. The number of CroCos is determined by the maximum number of CroCos that fit in a single layer within the cryostat. For the design of the CroCos, a *RE*BCO tape current per width of 30 A/mm at 77 K, s.f. is assumed to allow for a broad selection of manufacturers (Table 4.2). A Lift Factor of 1.5 is adopted since the highest temperature in the system is by design 70 K thus resulting in a current per width of 45 A/mm at 70 K s.f (s. Figure 4.9).

In order to achieve a current of 100 kA each of the 36 CroCos needs to be operated at about 2.8 kA and have a critical current of about 3.5 kA. Accordingly, critical current calculations were performed with the above described conditions by iterating the number of *RE*BCO tapes built into the CroCos until a solution was found. In Table 9.5 the results of the critical current calculations are shown for three manufacturers and two CroCos with different numbers of *RE*BCO tapes built in. Furthermore, the sum of critical currents of the individual tapes at 70 K, s.f. to gauge the critical current reduction due to the magnetic field of the cable is listed. The manufacturers were chosen because their *RE*BCO tapes offer distinctively different critical current angular magnetic field dependencies (see Table 4.3 and Figure 4.7). Depending on the critical current magnetic field dependence of the *RE*BCO tapes each of 36 CroCos needs between 18 and 20 6 mm wide tapes and 8 4 mm wide tapes.

To accommodate the high demand in stabilizer material the copper cross section can be increased to 19.2 mm² per CroCo without distorting the distinctive geometry of the conductor or decreasing the number of *RE*BCO tapes resulting in a total of 691 mm² of copper stabilizer built-into the CroCos. Inserting copper wires (2.3 mm and 2.7 mm diameter) into the gaps between CroCos provides another 355 mm² of copper stabilizer thus totaling 1046 mm² and exceeding the requirement.

Table 9.5: Individual critical current of CroCos of *RE*BCO tapes and different manufacturers

Number of *RE*BCO tapes		Tape I_c sum 70 K, s.f.	CroCo I_c 70 K, cable field SST	DNA	Theva
6 mm	20	6.8 kA	3.5 kA	3.7 kA	3.8 kA
4 mm	8				
6 mm	18	6.3 kA	3.3 kA	3.5 kA	3.5 kA
4 mm	8				

The losses of the superconducting cable amount to 1.25 W/m through the cryostat and the 0.1 W/m due to operating the cable close to the critical current.

The large cross section of the LN_2 channel is largely due to the restriction that only a single layer of CroCos is to be used. The cable could be significantly more compact at the cost of more superconductor material. The LN_2 mass flow rate in the coolant channel needs to be 0.141 kg/s to achieve an end temperature of 70 K at the given pressure drop.

The aim of cable design presented in this work is to provide cables that incorporate all necessary components into the cable including the stabilizer. For cables exceeding the 100 kA or significantly higher short circuit requirements it may be worth investigating if the stabilizer material could be removed from the superconducting cable and provided through a separate circuit in parallel. For example in order to achieve the short circuit requirements of the secondary

Table 9.6: Cable characteristic based on the design process

	Parameter	Value
CroCos	Numbers of CroCos	36
	6 mm wide *RE*BCO tapes	18 - 20
	4 mm wide *RE*BCO tapes	8
	Tape I_c sum (I_{c0}), 70 K, s.f.	6.3 - 6.8 kA
	I_c, 70 K, cable field	3.5 kA
	CroCo diameter	~7.5 mm
Stabilizer	Built-in stabilizer (CroCos)	691 mm²
	36 copper wires, ∅2.7 mm	206 mm²
	36 copper wires, ∅2.3 mm	150 mm²
	Total	1047 mm²
Losses	Cryostat losses	625 W at 67 K
	Superconductor losses	50 W at 67 K
	Total	675 W at 67 K
Coolant	Mass flow	0.141 kg/s
Dielectric	Cryostat dielectric, PPLP	1 mm

circuit cable ($5 \cdot I_{op}$ for 1 s) a 100 kA class cable would need about 3000 mm² of copper stabilizer which is about 3 times the amount built into the current conceptual design. Providing the remaining 2000 mm² would require a copper cable of roughly 50 mm diameter to be placed into the cable. This is illustrated by the dashed lines circle in Figure 9.2 and as can be seen requires a significant portion of the cable.

9.4 Secondary circuit

In following section a conceptual design for superconducting cable operating in the secondary circuit of an aluminum plant is outlined. The purpose of the secondary circuit is to provide a magnetic field that is capable of mitigating the magnetic field generated by the primary circuit. Therefore the secondary circuit is not strictly a power transmission cable but rather akin to a magnet.

A magnet design approach has been subsequently adapted by [Duv+14] for the secondary circuit of aluminum plants. The concept is to use multiple high current conductors (CroCos in this design study) and to connect them in series instead parallel. Due to the series connection the current now runs in multiple turns along cable. The advantage of this concept is that the current transferred to cryogenic temperatures can be drastically reduced and that current source can be down scaled. For example the 3.5 kA source used in Section 8.1.1 fits into a conventional 19 inch rack.

This conceptual design will therefore expand on the concept of a series connection of high current conductors.

Design specifications

The operational current of the secondary circuit is set to 100 kA to be identical to the primary circuit cable discussed in the previous section. Depending on the outline of the facility the secondary circuit may require between 10 % and 150 % of the primary circuit current [RUN95; Ren+16]. Operational current in this case means that the total current flowing over the cross section of the cable in one direction is 100 kA. This means that the CroCos in this scenario will be connected in series through 35 low resistive joints between the individual CroCos.

Since no significant voltage drop occurs within the superconductor, a voltage of lower then 50 V should be sufficient to operate the system.

In terms of fault resilience, the secondary circuit should not require sophisticated fault current limiters and therefore the the fault current is set to 5 times the operational current and the fault duration to 1 s.

The remaining design specifications are identical to primary circuit. For a detailed discussion refer to Section 9.3.

Table 9.7 summarizes the specifications for the secondary circuit cable.

Table 9.7: Design specifications and boundary conditions used in this case study

Description	Symbol	Value	Description	Symbol	
Total cross section current	I_{op}	100 kA	Inlet temperature	T_{in}	67 K
Operational voltage	U_{op}	<50 V	Outlet temperature	T_{out}	70 K
Cable length	L	500 m	Inlet pressure	p_{in}	1.5 MPa
Fault current	I_{fault}	$5 \cdot I_{op}$	Outlet pressure	p_{out}	1.499 MPa
Fault duration	t_{fault}	1 s	Operational to critical current ratio	$I_{\mathrm{op}}/I_{\mathrm{c}}$	0.8
Ambient temperature	T_{H}	293 K			
Inner cryostat diameter	d_{cryo}	130 mm			

Conceptual design

The conceptual cable design for secondary circuit as shown in Figure 9.3 is very similar to the primary circuit design (Figure 9.2) which is due to the large overlap in the specifications.

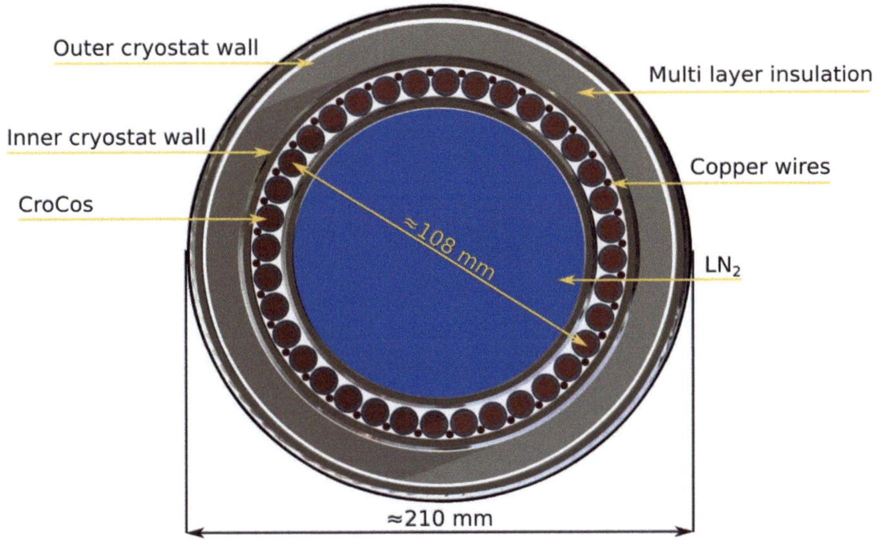

Figure 9.3: Schematic cross section of a 100 kA at 70 K consisting of 36 CroCos with an additional 36 copper wires to provide additional stabilizer material in case of a fault current.

The individual CroCo design is identical to the primary circuit, please refer to Section 9.3 for a detailed discussion of the CroCo design.

It may be necessary to insulate the CroCos from each other through a dedicated dielectric to ensure the series connection. However so called non-insulation magnets have gained increased attention over the last years [Hah+11; Yoo+16]. Non-Insulation magnets rely on the fact that the resistance along a the superconductor (including joints) is orders of magnitudes lower than any connections between the conductors and therefore forego any dedicated conductor dielectric. For the purpose of this work, it will be assumed that a layer 0.5 mm thick layer of PTFE can be extruded on the CroCos which should provide sufficient insulation if necessary.

The series connection of the CroCos inflicts additional losses due to the necessity of non-superconducting joints. Easy to manufacture joints between CroCos have achieved resistances in the range of 30-60 nΩ[WFP17]. The voltage drop across all 35 joints is therefore in between 3 mV and 6 mV resulting in up to 16 W at 67 K of additional losses for the cable system.

An advantage of the series connection is that the short circuit currents are significantly lower compared to the primary circuit. The expected fault current within the CroCo is 14 kA ($5 \cdot 2.8$ kA) for a duration of 1 s. The maximum permissible temperature rise during a fault is determined by the boiling temperature of the liquid nitrogen at a pressure of p_{out} = 1.499 MPa which is about 110 K (see Figure 6.1). Adding a safety margin of 2 K results in T_{end} = 108 K. Copper is used as stabilizer material and applying the code in Annex A.3 provides a stabilizer requirement of

Table 9.8: Cable characteristic based on the design process

	Parameter	Value
CroCos	Numbers of CroCos	36
	6 mm wide *RE*BCO tapes	18 - 20
	4 mm wide *RE*BCO tapes	8
	Tape I_c sum (I_{c0}), 70 K, s.f.	6.3 - 6.8 kA
	I_c, 70 K, cable field	3.5 kA
	CroCo diameter	~7.5 mm
Stabilizer	Built-in stabilizer	19 mm²
	36 copper wires, ⌀2.5 mm	176 mm²
	Total	195 mm²
Losses	Cryostat losses	625 W at 67 K
	Superconductor losses	50 W at 67 K
	Joints	16 W at 67 K
	Total	691 W at 67 K
Coolant	Mass flow	0.141 kg/s
Dielectric	CroCo dielectric, PTFE	0.5 mm

85 mm². The built-in copper stabilizer of 19.2 mm² in each CroCos is insufficient for this fault scenario. Additional copper wires have been added in Figure 9.3 to supply additional stabilizer material. The copper wires have a diameter of 2.5 mm and amount to a total cross section of 176 mm² which fulfills the requirements if they are connected in parallel.

The large cross section of the LN_2 channel that is largely due to the restriction that only a single layer of CroCos is to be used. The cable could be significantly more compact at the cost of more superconductor material. No return channel for the liquid nitrogen is included into the design. As can be seen in Figure 9.1 the cable start and end points are close together and therefore both ends can be directly connected to the refrigerator system making an dedicated return channel redundant. The LN_2 mass flow rate in the coolant channel needs to be 0.141 Kg/s to achieve an end temperature of 70 K at the given pressure drop.

9.5 Aluminum bus bar

A 100 kA bus bar system is outlined in this section as a reference system to the superconducting system previously discussed. Copper and aluminum are commonly used materials for bus bar systems in energy intensive industries due to their low resistance, mechanical properties, and price [OBH07, p. 728]. In case of an aluminum plant it will be assumed that aluminum bus bars are used.

Design specifications

The DIN 43670 standard describes the maximum operating current of aluminum bus bar systems based on a maximum operating temperature of the bus bar for currents up to about 10 kA. At 10 kA the operating current density is about 1.25 A/mm² with an operating temperature of 65 °C and ambient temperature of 35 °C. However, the current density can not be scaled up linearly with the current as the heat dissipation is reduced when bus bars are stacked [OBH07, p. 731]. For this case study a current density of j = 0.5 A/mm² and a operating temperature of 50 °C is assumed [Mor15]. The remaining design specifications are identical to the HTS system in order to maintain comparability.

Table 9.9: Design specifications and boundary conditions used in this case study

Description	Symbol	Value
Operating current	I_{op}	100 kA
Operating current density	j_{op}	0.5 A/mm²
Operational voltage	U_{op}	2 kV
Bus bar length	L	500 m
Fault current	I_{fault}	500 kA
Fault duration	t_{fault}	0.1 s
Ambient temperature	T_{H}	27 °C
Operating temperature	T_{op}	50 °C

Outline

The necessary aluminum cross section is calculated with the targeted operating current and the previously defined operating current density and is in this scenario 200000 mm². Due to cooling requirements this cross section can not be realized in a single bus bar, instead it needs to be separated into for example multiple plates to maximize the surface for air cooling. Joints to connect the individual bus bars will be neglected at this point.

The specific resistance of aluminum at 50 °C is $\rho = 3.17\,\text{n}\Omega\text{m}$ [OBH07, p. 732]. The losses per meter in a 100 kA bus bar system with the above mentioned values would then be

$$q_{Alu} = \rho j I = 3.17 \cdot 10^{-9}\,\Omega\text{m} \cdot 5 \cdot 10^{5}\,\text{A/m}^2 \cdot 10^{6}\,\text{A} = 1585\,\text{W/m} \tag{9.6}$$

9.6 System comparison

9.6.1 General properties

In Table 9.10 general properties of the above conceptually designed superconducting cable and aluminum bus bar are shown. It can be seen that the current density of the superconducting system is roughly six times higher than the aluminum bus bar system. The current density of the aluminum bus bar however neglects the minimum distance between the aluminum bars that are necessary for cooling. Including these distances would reduce the current density of the aluminum bus bar by roughly by a factor 2 to 3 [OBH07, p. 728f].

Table 9.10: Comparison of general properties of a superconducting cable and aluminum bus bar within the primary circuit of an aluminum smelter

	Primary circuit	Secondary circuit	Aluminum
Length	500 m	500 m	500 m
Operating current	100 kA	2.8 kA	100 kA
Current density	2.89 A/mm^2	2.89 A/mm^2	0.5 A/mm^2
Operating voltage	2 kV	50 V	2kV/50V
Critical current	≈125 kA	≈3.5 kA	-
Operating temperature	70 K	70 K	323 K
Operating pressure	1.5 MPa	1.5 MPa	-
Diameter	≈210 mm	≈210 mm	-
Cross section	0.035 m^2	0.035 m^2	0.2 m^2
Total weight	≈16 t	≈15 t	270 t
*RE*BCO tape, 6 mm	324 km - 360 km	324 km - 360 km	-
*RE*BCO tape, 4 mm	144 km	144 km	-

In terms of weight the superconducting system is 18 times lighter than the aluminum system. In addition significant savings in terms of support structure can be expected with the superconducting cable only weighting about 30 kg/m compared to the 540 kg/m of the aluminum bus bar.

A detailed comparison of the overall system losses, operating and investment cost is performed in the following sections.

9.6.2 Annual losses and operating cost

To judge the efficiency of superconducting systems it is important to not only know the losses at low temperature but also estimate the amount of energy that is required to compensate these losses. For this, the so called coefficient of performance (COP) is often used. It is defined as the ratio of the heat removed at cryogenic temperature to the amount of work that must be applied to remove it.

$$COP = \frac{Q}{P} \tag{9.7}$$

The minimal amount that is required to cool down can be calculated with the ideal Carnot refrigerator cycle

$$COP_{Carnot} = \frac{T_{cold}}{T_{ambient} - T_{cold}} \tag{9.8}$$

with T_{cold} being the cold temperature and $T_{ambient}$ the ambient temperature in K. Since real refrigerators can not achieve the efficiency of the Carnot process the so called 'Figure of Merit' (FOM) or Carnot-Efficiency is used to compare real refrigerators to the ideal process.

$$FOM = \frac{COP}{COP_{Carnot}} \tag{9.9}$$

Losses

Table 9.11 lists the losses in the current leads and the cable for the two discussed superconducting uses cases and a reference aluminum bus bar. For this case study a two-stage current lead was chosen based on a study that investigated two- and three-stage current leads for 64 K operation and an ambient temperature of 293 K [Bro+]. The three stage current lead showed a reduction in losses by about 6 % compared to the two stage current lead. The operation temperature in [Bro+] is close enough to the lowest temperature within this case study (67 K) so that the heat loads will be directly adopted. The heat loads at the two stages are then 39 W/kA at 145 K and 17 W/kA at 67 K. For detailed conceptual designs for current leads please refer to the literature [Gol04; Sch+17; Bro+].

Table 9.11: Loss summary for primary, secondary circuit superconducting cables and aluminum bus bar assuming an ambient temperature of 293 K.

	Current lead / W	Cable / W	P_{ideal} / W	ΣP_{ideal} / W	P_{real} / W	ΣP_{real} / kW	P_{annual} / MWh/a
Primary circuit	7800 at 145 K	-	7961		39807	112	977
	3400 at 67 K	675 at 67 K	13746		72882		
Secondary circuit	272 at 145 K	-	279	3011	1858	20	175
	119 at 67 K	691 at 67 K	2732		18215		
Aluminum bar	-	-	-	-	-	791	6877

The cable losses in Table 9.11 consist for largest part of cryostat heat leakage (625 W) and the losses of the superconductor due to being operated close to the critical current (50 W). Losses introduced by circulation pumps are negligible in this case due to the low pressure drop. In case of the secondary circuit an additional 16 W of losses are introduced by the joints necessary to connect the CroCos in series.

The ideal power consumption P_{ideal} at 293 K is calculated by using the coefficient of performance of the ideal Carnot process (equation 9.8) and are equal to

$$COP_{Carnot}\,(145\,\mathrm{K}) = \frac{145\,\mathrm{K}}{293\,\mathrm{K} - 145\,\mathrm{K}} = 0.98 \tag{9.10}$$

$$COP_{Carnot}\,(67\,\mathrm{K}) = \frac{67\,\mathrm{K}}{293\,\mathrm{K} - 67\,\mathrm{K}} = 0.296 \tag{9.11}$$

For an estimation of the real power consumption P_{real} a Figure of Merit of 0.2 for primary circuit and 0.15 for the secondary was assumed [Gre15]. The difference is introduced to factor in that refrigerators become more efficient at higher capacities [Str; Rad09; Gre15].

An estimation of the real power consumption P_{real} at 293 K is calculated by applying a Figure of Merit of 0.2 for primary circuit and 0.15 for the secondary circuit cable to the ideal power consumption [Gre15].

$$P_{real} = \frac{P_{ideal}}{FOM} \tag{9.12}$$

The annual losses P_{annual} are calculated by

$$P_{annual} = t_{annual} \cdot \Sigma P_{real} \tag{9.13}$$

with t_{annual} being the operational time per year. In this case study nearly non-stop operation of t_{annual} = 8700 h/a is assumed.

Operating cost

The electricity price for non-household consumers with a consumption between 70 GWh and 150 GWh was 45.7 €/MWh in Germany and 58.2 €/MWh on average in the European Union excluding taxes and other levies [eur19]. Large consumers that considerably exceed the 150 GWh can have prices significantly lower and are in the range of 30 €/MWh but even for these industrial consumer electricity prices have been rising in recent years [Ask18]

Table 9.12 shows the annual electricity cost accrued by operating the superconducting cables discussed in the case study and a reference aluminum bus bar for the three above mentioned electricity prices. Within the primary circuit the electricity cost can be reduced by about 85 % compared to the conventional aluminum bus bar. In the secondary circuit the cost is reduced even further by about 97.5 % amounting the several hundred thousand Euro per year.

Table 9.12: Comparison of operating cost of conventional and superconducting solutions for three different electricity prices

	Annual cost of electricity		
Electricity price	30 € / MWh	45.7 € / MWh	58.2 € / MWh
Primary circuit	29 k€ / a	45 k€ / a	57 k€ / a
Secondary circuit	5 k€ / a	8 k€ / a	10 k€ / a
Aluminum bar	206 k€ / a	314 k€ / a	400 k€ / a

The maintenance cost are neglected at this point as they are strongly dependent on the chosen cryogenic system for the superconducting system.

9.6.3 Investment cost

In the following the investment cost of the two superconducting cable systems and the aluminum reference system will be discussed. For each of the systems, a high and low cost estimation will given. The investment cost are based on offers from manufacturers when possible.

Table 9.13 and 9.14 show two investment cost scenarios for a superconducting cable system. In general it can be noted that system cost of a superconducting cable is dominated by the cost of the superconductor that account for roughly 65 % of the total system cost. The high variance in the system cost is also due to the high variance in the superconductor cost. The high cost scenario of the superconductor represents the best offer received during the demonstrator cable design phase while the low cost considers a price drop by 50 %. The large price drop appears to be reasonable since advances in manufacturing and a discount for large orders can be factored in. In addition it was shown in Figure 8.4 that the average critical current of tapes is a good foundation for the calculation of the critical current of a CroCo. This potentially allows for less strict tape specifications on the minimum critical current of the superconductor which could also lead to cost reductions. The price of the copper tapes within the CroCo is based upon London Metal Exchanges rates. The CroCo solder price is based upon orders performed by the author.

The superconducting cable manufacturing cost are based on the experience of CroCo manufacturing performed in this work.

Flexible cryostat sizes in the area of DN 130 and larger are non standard products and therefore the price was extrapolated from offers of smaller flexible cryostats.

The refrigerator cost is estimated based on a study performed in [Gre15]. The study showed that there is significant variance at the capital of cryocooler is for the LN_2 region. Therefore the following function gives only a rough estimate

$$C = 1.81 \cdot R^{0.57} \tag{9.14}$$

Table 9.13: Investment cost of a 500 m superconducting cable within the primary circuit

Component	Low cost in k€	High cost in k€	Comments
Superconducting Cable	**5507**	**10611**	
CroCo	4256	8361	*RE*BCO tape 50 - 100 €/(kA m), copper 7.3 €/kg, SnAg4 solder 50 €/kg
Cryostat	750	1500	Flexible cryostat e.g. Nexans Cryoflex, DN130, 1500 - 3000 €/m
Manufacturing	500	750	1 cm/s CroCo production speed, 60 €/h labor cost
Cryogenic system	**800**	**1180**	
Refrigerators	250	280	5 kW at 67 K
Current leads	400	600	Two-stage current leads
Accessories	150	300	Control system, pipes, pumps, etc
Installation	50	100	Estimation based on [Rei14]
Total system cost	**6357**	**11891**	

C being the capital cost in thousand US$ and the R the refrigerator power in W at 77 K. The cooling capacity of cryocoolers declines as the temperature decreases. The decline is usually provided by the manufacturers in so called capacity maps. In Section A.7 the capacity maps of commercially available cryocoolers (Table 9.1) were considered to calculate the capital cost at 67 K. This results in capital cost equal to

$$C = 1.81 \cdot \left(\frac{R}{a \cdot T + b} \right)^{0.57} \tag{9.15}$$

with T being the temperature in K and a = 0.009 1/K, b = 0.303 in the low cost scenario and a = 0.024, b = -0.866 in the high cost scenario (see Table A.2). The losses are about 4000 W at 67 K and 800 W at 67 K for the primary and secondary circuit respectively. For the capital cost of calculation a cold power capacity buffer of 25 % was set and therefore 5 kW at 67 K for the primary circuit cable and 1 kW at 67 K for the secondary circuit cable was used.

The installation cost of the secondary circuit is slightly more expensive due to the necessity of manufacturing the joints between the CroCos on site.

Table 9.15 shows the investment cost of an conventional 100 kA aluminum bus bar. A major part of the investment cost is installation which includes support materials for the roughly 270 t of aluminum.

Table 9.14: Investment cost of a 500 m superconducting cable within the secondary circuit

Component	Low cost in k€	High cost in k€	Comments
Superconducting Cable	**5474**	**10578**	
CroCo	4224	8328	*RE*BCO tape 50 - 100 €/(kA m), copper 7.3 €/kg, SnAg4 solder 50 €/kg
Cryostat	750	1500	Flexible cryostat e.g. Nexans Cryoflex, DN130, 1500 - 3000 €/m
Manufacturing	500	750	1 cm/s CroCo production speed, 60 €/h labor cost
Cryogenic system	**400**	**610**	
Refrigerators	100	110	1 kW at 65 K
Current leads	200	300	Two-stage current leads
Accessories	100	200	Control system, pipes, pumps, etc
Installation	100	200	Estimation based on [Rei14] and slightly increased to joint manufacturing
Total system cost	**5974**	**11388**	

In comparison, it can be seen that the investment cost of the superconducting system are roughly 3 to 5 times higher compared to a conventional aluminum bus bar.

Table 9.15: Investment cost of 500 m of aluminum bus bar

Component	Low cost in k€	High cost in k€	Comments
Aluminum bus bar	**1673**	**2447**	
Material	561	758	Material cost between 1850 $/t and 2500 $/t [The19]
Manufacturing	112	189	Assumption: Manufacturing is between 20 % and 25 % of material cost
Installation	1000	1500	Estimation based on [Rei14]

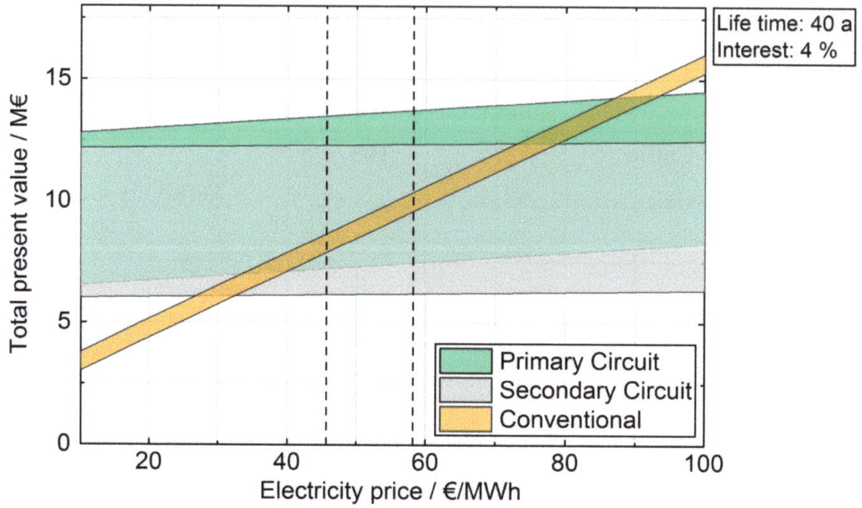

Figure 9.4: Total present value of superconducting cable systems within the primary and secondary circuit at an operational current of 100 kA over 500 m compared to a conventional aluminum bus bar system for a life time of 40 years and 4 % interest.

Present value as a function of electricity price

The present value offers the possibility to evaluate all cost of an investment (for example operating cost) by calculating their value at a given time. The operating cost in this scenario will be discounted over the life time of the system with a fixed interest rate. Assuming the operating cost are paid at the end of the year and the interest rate r is constant then the present value C can be calculated by

$$C = I + O\frac{(1+r)^N - 1}{(1+r)^N \cdot r} \tag{9.16}$$

I being the initial investment cost, O the operating cost and N the operating life time [TBE19, p. 80f].

The total present value of the three systems discussed in this chapter are shown as a function of the electricity price in Figure 9.4. The colored areas shows the spread of the total present value due to the high and low investment cost estimations performed in this section (Tables 9.13,9.14, &9.15). The dashed vertical lines represent the average electricity prices of consumers with up to 150 GWh consumption in Germany (also see Table 9.12). An operational life time of 40 years and an interest rate of 4 % was assumed for the displayed figure.

The total present value of electricity cost in the secondary circuit system are only about 350 k€ at 100 €/MWh and therefore the total present value of the secondary circuit (Figure 9.4)

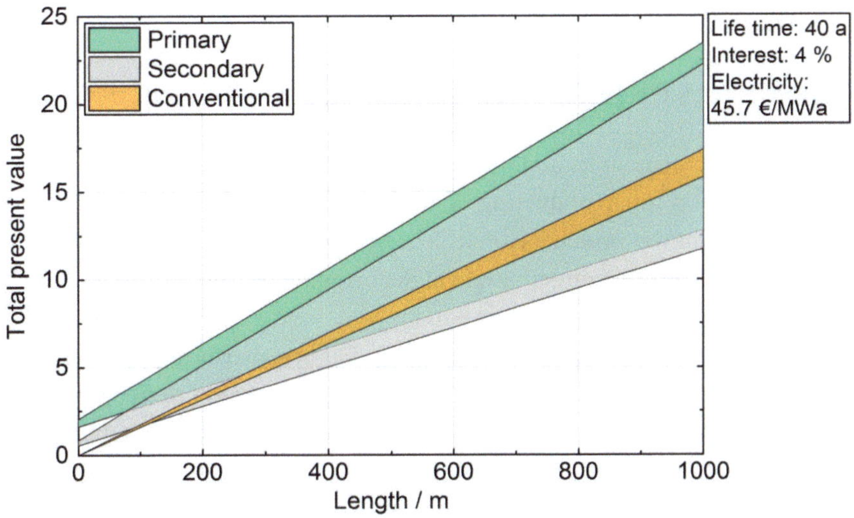

Figure 9.5: Total present value of superconducting 100 kA cable in the primary and secondary circuit compared to a conventional aluminum bus bar as a function of cable length

appears to be roughly independent of the electricity price. The present value of the conventional system increases steadily and shows the first the first overlap with a superconducting system at electricity prices around 30 €/MWh with the secondary circuit system and at around 35 €/MWh with the primary circuit system. This means that under ideal conditions, superconducting system could be competitive investment alternatives to conventional solutions for electricity prices above 35 €/MWh.

Present value as a function of cable length

In the previous sections a length of 500 m was assumed in all solutions to maintain comparability. To address cables of different lengths Figure 9.5 shows the total present value as a function of cable length at an electricity price of 45.7 €/MWh [eur19], a life time of 40 years and an interest rate of 4 %. For the calculation the operational and investment cost of the previously outlined cable concepts were expressed as a function of the cable length. This means that additional technical challenges for long cables such as increased pressure drop were neglected at this point.

In Figure 9.5 it can be seen that the break even point between a superconducting and a conventional solution occurs around 100 m and 300 m for the low cost assumption of the secondary and primary circuit respectively. For the high cost assumptions it can be seen that no beak even point will be achieved with the current parameter set since the discounted the cost per meter of the superconducting system exceeds the conventional systems.

9.7 Chapter summary

In this chapter the application of a high current HTS cable system in an aluminum smelting plant has been investigated. A use case of a superconducting cable within the primary circuit between HVDC rectifiers units and the electrolysis pot has been identified. A second use case is the adaptation of a superconducting system in a secondary circuit that is used to mitigate the magnetic field of the primary circuit. The second use case is particularly interesting for superconducting cables utilizing multiple high current conductors such as the cable designed, build, and tested in this work (Chapters 6 to 8).

A conceptual design for a superconducting cable has been performed for each of the identified uses cases with a rated operational current of 100 kA. As a reference a conventional aluminum bus bar system as been conceptually designed. The losses of all three systems were calculated based on an analysis of current lead concepts and cryogenic system concepts. An analysis of the operating cost showed that a superconducting system could save several hundred thousand Euros per year.

The investment cost of each of the three conceptually designed system was assessed by considering a low and high investment cost scenario. In a subsequent analysis of the total present value of the investments it could be shown that superconducting system could be competitive choices at electricity prices around 30 €/MWh to 35 €/MWh.

10 Summary and outlook

The focus of this thesis was the development, manufacturing and testing of the superconducting elements of a 35 kA HTS DC cable demonstrator. Three major HTS high current conductor concepts (Stack, Roebel, cable on round core (CORC)) were analyzed and the HTS Cross Conductor (HTS CroCo) stack concept was chosen. A design process for HTS DC cables was developed for high current applications (Chapter 6). Based on the design process, a 100 kA cable design based on the 36 CroCos was developed and subsequently down scaled to a 35 kA cable demonstrator utilizing 12 CroCos. The length of the cable demonstrator was defined as 3.6 m and the operating temperature was set to 77 K (Section 6.8).

Prior to the CroCo manufacturing, a study of the critical current degradation behavior of second generation HTS tapes under thermal loads was performed (Section 7.1). These investigations are necessary since *RE*BCO tapes are often exposed to thermal loads during manufacturing for example due to soldering. An understanding of the precise degradation behavior can therefore help to optimize the manufacturing processes. The investigation found a predictable and repeatable degradation pattern that appears to be caused by oxygen depletion of the *RE*BCO crystal. It was shown that tin-silver solders could be a viable low-cost alternative to lead containing solders in CroCo manufacturing. The results of the investigation are published in [Pre+18].

Towards the manufacturing of the CroCos for the cable demonstrator, *RE*BCO tapes from two manufacturers were investigated in regards to their compatibility with the CroCo manufacturing process (Section 7.3). Tapes from one manufacturer were found not compatible at the time but are promising candidates for future work. It was found that *RE*BCO tapes from other manufacturer were in principle compatible but had strong tendencies towards an unequal distribution of electroplated copper. However this problem was not critical for the formation of the CroCos and the tapes could be used.

The CroCo manufacturing process was further developed to incorporate a round outer solder matrix. In addition, the manufacturing process was adapted to accommodate *RE*BCO tapes with minimal electroplated copper (≈5-10 μm) which helps to reduce the cost of the *RE*BCO tapes. To enable these major changes a new CroCo manufacturing machine was set up (Section 7.2). The 12 CroCos used in the demonstrator cable were produced in 8 production runs. From each CroCo production run, one or two of the desired 3.6 m long CroCos were cut. The critical current of each of the 12 CroCos were measured under self field conditions at 77 K and varied between 2890 A and about 3680 A. The relatively large variation is due to optimizing various manufacturing steps and an increasing performance of the *RE*BCO tapes. It was shown that the critical current of a CroCo can be reliably calculated based on the average critical current of *RE*BCO tapes used to manufacture the CroCo (Section 8.1.1). Soldered joints were used to connect the CroCos with normal conducting copper cables. The resistance of these joints was

200 nΩ on average with a standard deviation of 43 nΩ at 77 K. This results in a heat load about 0.7 mW per connection at an operational current of 3500 A which should be sufficiently low for most applications.

In order to test the cable demonstrator, a 6 m long, 1 m wide and 0.8 m high cryostat was developed and built (Section 8.2.1). The demonstrator cable itself consists of the 12 CroCos, each incorporating 22 6 mm wide and 10 4 mm wide *RE*BCO tapes. The CroCos are mounted upon an aluminum core with a diameter of 110 mm and connected in parallel (Section 8.2.2).

During the measurement series the cable demonstrator achieved a steady state current of 34 kA and 35 kA for a short period of time. An analysis of the current distribution during the cable test showed a relatively large spread in the current distribution across the CroCos that was likely caused by water condensing between contacts and thus increasing the contact resistance (Section 8.3.2). A CroCo was dismounted from the demonstrator after the test to verify that no degradation for example due to strain during the cool down occurred (Section 8.3.3).

This thesis concludes with a case study of a superconducting cable in an aluminum electrolysis plant. It is shown that the energy savings for a 500 m long 100 kA class superconducting cable compared to a conventional aluminum bus bar are in the range of 6 GWh to 6.5 GWh per year which is roughly equal to the yearly energy consumption of 2000 two person households in Germany. The investment cost for the above mentioned superconducting system is presently three to six times higher than a conventional aluminum bus bar system. The large spread in the investment cost of the superconducting system is largely due to a large variance in the superconductor cost that dominate the investment cost. The break even point between the superconducting system and the conventional system for the minimal investment cost of both systems is around an electricity price of 35 €/MWh calculated over a period of 40 years.

In conclusion, this thesis performed a first step towards demonstrating the technical feasibility of superconducting high current DC cables and showed an economically feasible use case in the foreseeable future.

The time constraints of this work allowed only for a brief measurement series of the 35 kA cable demonstrator. A series of experiments should be performed in future work to further investigate the capabilities of a high current cable based on multiple CroCos. A first step is the improvement of the current distribution across all CroCos by ensuring more homogeneous contact resistances. A more compact aluminum core could facilitate a more compact cable in which the magnetic field of the cable has a larger impact compared to the magnetic field of the individual CroCo. In addition, the behavior of the demonstrator during transients for example short circuits should be investigated for example in regards to heat development.

A Appendix

A.1 Order of magnitude estimation of the coolant friction

In the temperature rise there are two terms discussed one is the external heat load due to heat leak through the cryostat or dissipation within the superconductor. The other factor is the coolant friction during the flow. The order of magnitude of the impact of the friction factor is quickly estimated

$$\Delta T = \frac{\zeta \cdot v^2 \cdot L}{2 \cdot D_h \cdot c_p} + \frac{q \cdot L}{\dot{m} \cdot c_p} \tag{A.1}$$

The AmpaCity cable which has a mass flow of 0.425 Kg/s, is 1000 m long and the diameter of the supply line is 32 mm.

$$\Delta T = \frac{\zeta \left(\frac{\dot{m}}{\rho_{LN2}\,\pi\, r^2}\right)^2 L}{4\, r\, c_{p,LN2}} = \frac{\zeta\, \dot{m}^2\, L}{4\, \pi^2\, r^5\, c_{p,LN2}\, \rho_{LN2}^2} \tag{A.2}$$

Assuming a friction factor of 0.02 results

$$\Delta T = \frac{0,02 \cdot 0,425 \cdot 1000}{4 \cdot \pi^2 \cdot 0,016^5 \cdot 2000 \cdot 807^2} \mathrm{K} = 0.067\,\mathrm{K} \tag{A.3}$$

A.2 Relevant copper material properties for the use as stabilizer material

The heat capacity of copper between 50 K and 400 K was fitted an exponential function based on data from [WC84]

$$c(T) = a_0 + a_1 \exp\left(a_3 T\right) \tag{A.4}$$

with $a_0 = 3.513 \cdot 10^6\,\frac{\mathrm{J}}{\mathrm{m^3K}}$, $a_1 = -5.414 \cdot 10^6\,\frac{\mathrm{J}}{\mathrm{m^3K}}$ and $a_2 = -0.014\,\frac{1}{\mathrm{K}}$.

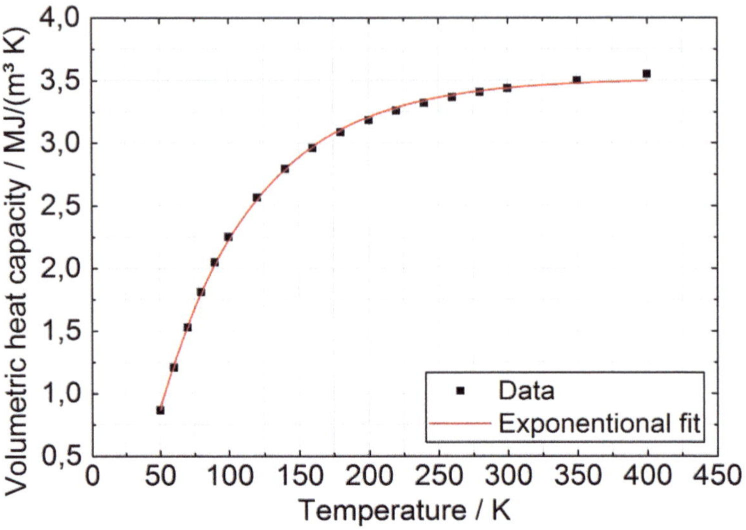

Figure A.1: Temperature dependent volumetric heat capacity of pure copper between 50 K and 400 K

A.3 Temperature incremental code to calculate stabilizer cross section

Code written in MATLAB to calculate the stabilizer cross section.

```
clear all
I_fault = 50000; % fault current

T_start = 77; % start temperature
T_end = 373; % end temperature
t_fault = 1; % duration

%% Volumetric heat capacity of copper 50 < T < 400 K,
% cp(T)=a0+a1*exp(a3*T) [J/(m^3 K)]
a0 = 3.51329E6;
a1 = -5.41352E6;
a2 = -0.01436;

%% Eletrical resistance
% Electrical resistance of copper 50 < T < 400 K,
% rho_cu(T) = c0+c1*T [Ohm*m]
```

```
c0  =-3.162*1E-9;
c1  = 6.798*1E-11;

%% Calculations
% Temperature increment
T_step = 0.1;
% List of temperature increments
T_list = T_start:T_step:T_end;
A = 0; % cross section in m^2
total_time = 0;
% until the time to T_end is higher than t_fault
while total_time < t_fault
A = A+1E-6; % add one mm^2 per loop through
run_1 = 1; % reset of run variable

for ii=1:size(T_list,2)-1 % loop through
% integrated heat capacity of the temperature increment...
int_cp(run_1) = (a0*(T_list(ii+1)-T_list(ii))+...
a1/a2*(exp(T_list(ii+1)*a2)-exp(T_list(ii)*a2)))*A;
% total power dissipation
heat(run_1) = I_fault^2*(c0+c1*T_list(ii+1))/A;
% time to reach the T_end of the temperature increment
t(run_1) = int_cp(run_1)/heat(run_1);
run_1 = run_1 +1;
total_time_stamp(run_1) = sum(t);
end
% total fault duration
total_time = sum(t);
clear int_cp heat t
end

A_inc = A*1E6;
%% Worst case
rho_start = c0+c1*T_start;
rho_end = c0+c1*T_end;

int_cp = (a0*(T_end-T_start)+a1/a2*(exp(T_end*a2)...
                -exp(T_start*a2)));
```

```
heat = I_fault^2*(c0+c1*T_end)*t_fault;

A_wc = sqrt(heat/int_cp)*1E6;

%% Displaying values
disp(['Incremental A: ', num2str(A_inc), ' mm^2'])
disp(['Worst case A: ', num2str(A_wc), ' mm^2'])
plot(total_time_stamp, T_list)
```

A.4 Darcy friction factor for smooth and corrugated pipes

Smooth and rough pipes

For a pipe the Darcy friction factor depends on the relative roughness of the pipe ϵ_ζ/D_h and the Reynolds number Re [RRM02]. In laminar flow the friction factor can be determined by the Hagen-Poiseuille equation

$$\zeta = \frac{64}{Re} \tag{A.5}$$

For turbulent flow with Reynolds numbers between 4000 and $10^8$and relative roughness from 0 to 0.05 the friction factor can be estimated with the Colebrook-White equation

$$\frac{1}{\sqrt{\zeta}} = -2\log_{10}\left(\frac{\epsilon_\zeta/D_h}{3.71} + \frac{2.52}{Re\sqrt{\zeta}}\right) \tag{A.6}$$

Eq.A.6 being implicit it has to be solved numerically. However there are several explicit approximation of the Colebrook-White equation. [RRM02] compared several of such approximations and concluded that eq A.7 provides, statistically, the best approximation

$$\begin{aligned}\frac{1}{\sqrt{\zeta}} &= -2\log_{10}\left[\frac{\epsilon_\zeta/D_h}{3,7065} - \frac{5,0272}{Re}\cdot\right. \\ &\left.\log_{10}\left(\frac{\epsilon_\zeta/D_h}{3,827} - \frac{4,567}{Re}\log_{10}\left(\left(\frac{\epsilon_\zeta/D_h}{7,7918}\right)^{0,9924} + \left(\frac{5,3326}{208,815+Re}\right)^{0,9345}\right)\right)\right]\end{aligned} \tag{A.7}$$

for $3000 < Re < 1.5 * 10^8$ and $0 \leq \epsilon_\zeta/D_h \leq 0.05$.

These formulas utilize the hydraulic diameter so they can be used for the ring as well as the normal pipe geometry.

Corrugated Pipes

The meta study of [Ji+15] shows that the friction factor of various corrugated pipes is 1.5 to 4.5 times higher than the friction factor of a smooth pipe with the same diameter. However it also shows that no overarching empirical formula has been found that is valid for a large range of diameters or Reynolds numbers. Even with the multitude of studies listed in the meta study there are still a lot of cases that aren't covered by the empirical formulas found in these studies. This leaves only the experimental determination of the friction factor in a lot of cases.

One empirical formula that was not listed in the meta study was determined by Knut Kauder [Kau71]. Since his PhD thesis is not readily available his empirical formula is repeated here

$$\zeta = 3400 \cdot \sqrt[10]{\left(\frac{e}{d}\right)^{6}\left(\frac{p_{cor}}{e}\right)^{7}}\left(\frac{e}{d}\right)^{3,53}\left(\frac{e}{p_{cor}}\right)^{230\cdot\left(\frac{e}{d}\right)^{2,1}} Re^{0,193\cdot exp\left(-3300\left(\frac{e}{d}\right)^{2,6}\right)\frac{e}{p_{cor}}} \tag{A.8}$$

within the borders of $0.2 \leq e/p_{cor} \leq 0.6$, $5 \cdot 10^4 \leq Re \leq 3 \cdot 10^5$ and $0.0455 \leq e/d \leq 0.0635$ with d being the inner diameter of the corrugated pipe, p_{cor} the pitch and the groove depth e. According to [Kau71] the transition to smooth pipe behavior starts at $e/p_{\mathrm{cor}} < 0{,}2$ so Eq A.7 with $\epsilon = 0$ can be used for these cases.

A.5 Critical current calculation of manufactured CroCos

In Figure 8.4 of section 8.1.1 calculated critical currents are compared with measured critical current of all manufactured CroCos.

Three calculations were performed based on different initial values for the critical current of the *RE*BCO tapes. The initial values are based on the critical currents of the batches used to manufacture a CroCo. Figure A.2 shows schematically the critical current variation of a *RE*BCO tape along its length of two different batches. It can be seen that average critical current in Batch 2 is higher than in Batch 1 and that the deviation is lower. The manufacturer often provides information similar to Figure A.2 for each batch. In addition to that, minimum, maximum, and average critical currents of the batch are determined by the manufacturer and provided to the customer.

In the CroCo manufacturing each CroCo consisted of 22 6 mm wide tapes and 10 4 mm wide tapes of several different batches. For example Table A.1 shows the batches of the 6 mm wide *RE*BCO tapes and their average, maximum and minimum critical current that were used to manufacture CroCo 8. The number of tapes used of a respective batch was multiplied with the three different critical currents and the resulting critical currents were than averaged across their respective category (Avg I_{c} , Max I_{c} , Min I_{c}). The final average per tape critical current were than used as initial values for critical current calculations that are shown in Figure 8.4.

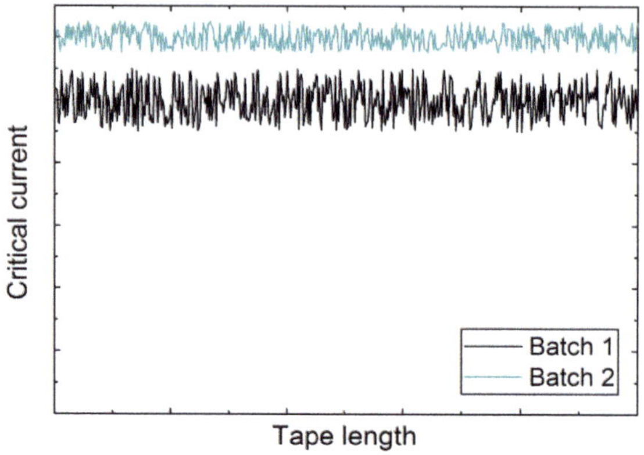

Figure A.2: Critical current variation of *RE*BCO tapes along their length

The same process was applied to the 4 mm wide tapes.

Table A.1: Determination of initial critical current values of 6 mm wide tapes in CroCo 8

	Batch			CroCo			
Number	Avg I_c / A	Max I_c / A	Min I_c / A	Tapes	Avg I_c / A	Max I_c / A	Min I_c / A
ST1707-19	203	224	185	1	203	224	185
ST1707-18	196	219	180	4	784	876	720
ST1707-17	195	211	180	1	195	211	180
ST1712-6	215	247	180	1	215	247	180
ST1712-7	210	230	186	2	420	460	372
ST1712-8	259	290	209	1	259	290	209
ST1712-9	251	281	207	1	251	281	207
ST1712-10	245	279	197	1	245	279	197
ST1712-11	252	274	208	3	756	822	624
ST1712-12	245	276	206	1	245	276	206
ST1712-13	246	298	186	4	984	1192	744
ST1712-14	263	300	228	1	263	300	228
Average per tape					228	257	192

The critical current calculations were performed with a FEM model within the commercial software COMSOL. Figure A.3 shows geometric model that was used for the CroCos and the magnetic field distribution of the CroCo at its critical current. As can be seen the geometric

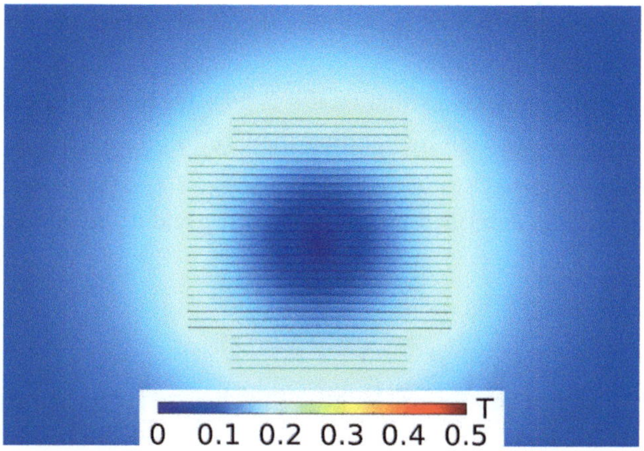

Figure A.3: Magnetic field distribution of CroCo 8 at its critical current

model consist only of thin rectangles with a thickness of 1 µm which model the *RE*BCO layer of the *RE*BCO tape. All other components of the CroCo were neglected for the critical current calculation. The electrical current within the *RE*BCO layers was imposed as an external current with a magnetic field dependence in accordance to the measured angular dependency of the used tapes (see Figure 4.7 and Table 4.3). The critical current magnetic field dependence was scaled with critical currents as determined by the process described above.

The SuperPower tapes used to manufacture CroCo 11 and 12 followed the advanced pinning formula which results in the very complicated angular dependency. In [Par+11] a model was proposed based on equation 4.4 that requires 14 manually determined parameters to actuartely represent the critical current angular magnetic field dependency of a SuperPower tape with advanced pinning.

Instead of choosing the complicated accurate model described in [Par+11] it was decided to simplify equation 4.4 to a magnetic field dependence by averaging the critical current across all angles resulting in the follow equation

$$j_e(B) = \frac{j_{e0}}{[1 + B/B_k]^{k_2}} \tag{A.9}$$

A.6 Thermal contraction of superconducting cables

The cold parts of a superconducting cable will contract during the cooldown with respect to the outer (warm) cryostat wall. The radial contraction is due to the small radial dimensions of cables of neglectable but the axial contraction can amount to several meters in long superconducting cables. Inhibiting the contraction can lead to significant strain and possibly damage to the cable (see Section 7.4.3). Another possibility is to compensate for axial contraction for example with movable cable terminations [ST04] or to translate the axial movement due to the contraction into a radial movement.

In the following it will be quickly estimated if it is possible to translate the axial contraction of a CroCo into a radial contraction by bending the CroCo into a helix at the example of the demonstrator cable.

In section 5.2.1 the helix is described by its twist pitch p and its helix radius r. The arc length of the helix is then

$$s = \sqrt{p^2 + (2\pi r)^2} \tag{A.10}$$

Due to the thermal contraction the arc length of the helix changes during cooldown.

$$s \cdot (1 + \Delta T \cdot \mathrm{CTE}) = \sqrt{p^2 + (2\pi r)^2} \tag{A.11}$$

The new helix radius is then

$$r_2 = \frac{\sqrt{(s \cdot (1 + \Delta T \cdot \mathrm{CTE}))^2 - p^2}}{2\pi} \tag{A.12}$$

Considering a thermal expansion coefficient of $11 \cdot 10^{-6}\ 1/\mathrm{K}$, a twist pitch of p = 1.5 m and an initial helix radius of r = 58 mm. Than the new radius would be r_2 = 55 mm. Allowing for 3 mm of radial movement of the CroCo would be sufficient to translate the complete axial contraction into a radial and thus not require any additional compensation for example at the terminations.

A.7 Cryocooler capacity maps

In section 9.6.3, equation 9.14 a cost function is given for cold power at 77 K. In order to estimate the capital cost at 65 K the capacity maps of the cryocoolers listed in Table 9.1 are shown in Figure A.4. For comparison the cold power of cryocoolers is normalized to the 77 K value. It can be noted that the decline between 77 K and 65 K is roughly linear for all cryocoolers and that the magnitude of the decline varies.

Assuming a linear function such as

$$P = T \cdot a + b \tag{A.13}$$

than the following parameters can be found for the cryocoolers shown in Table A.4.

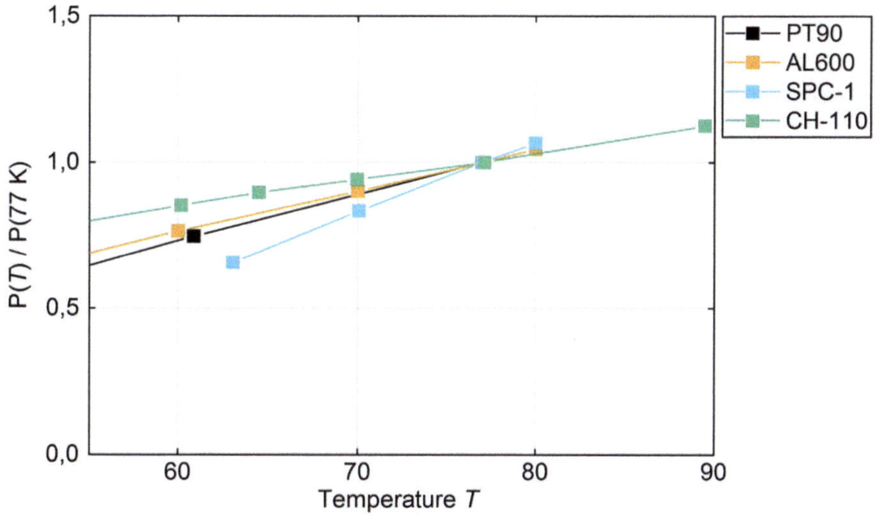

Figure A.4: Cold power as a function of temperature normalized to the cold power at 77 K of commercially available cryocoolers (based on [Cry19; Sti19; SHI19])

Table A.2: Linear fit parameters for the normalized cold power of cryocoolers as a function of temperature

	a / 1/K	b
PT90	0.016	-0.204
AL600	0.014	-0.07
SPC-1	0.024	-0.866
CH-110	0.009	0.303

B List of abbreviations

AC	Alternating current
ACT	Advanced Conductor Technologies
AMSC	American Superconductor
Bi-2212	Bi2Sr2CuO6+δ
Bi-2223	Bi2Sr2Ca2Cu3O10+δ
CAD	Computer aided design
CC	Grooved copper cylinder
CERN	European Organization for Nuclear Research
CL	Copper litz
COP	Coefficient of performance
CORC	Conductor on Round Core
CroCo	HTS Cross Conductor
6/4 CroCo	CroCo using tape width of 6 mm and 4 mm
4/2 CroCo	CroCo using tape width of 4 mm and 2 mm
3/2 CroCo	CroCo using tape width of 3 mm and 2 mm
CS	Cable shoe
CTE	Coefficient of thermal expansion
DC	Direct current
DNA	Deutsche Nanoschicht
DIN	Deutsches Institut für Normung
ENEA	Agenzia nazionale per le nuove tecnologie, l'energia e lo sviluppo economico sostenibile
FEM	Finite element method
FOM	Figure of Merit
GCS	General Cable Superconductors
GFRP	Glass fiber reinforced plastic
H_2S	Hydrogen sulphide

HTS	High temperature superconductor(s)
IBAD	Ion beam assisted deposition
IEC	International Electrotechnical Commission
IEEE	Institute of Electrical and Electronics Engineers
ISD	Inclined substrate deposition
KIT	Karlsruhe Insitute for Technology
LaH_{10}	Lanthanum hydride
LF	Lift factor
LHC	Large Hadron Collider
LN_2	Liquid nitrogen
LTS	Low temperature superconductor(s)
MgB_2	Magnesium diboride
MOCVD	Metal-organic chemical vapour deposition
MOD	Metal-organic deposition
MIT	Massachusetts Institute of Technology
MLI	Multi-layer super insulation
NbTi	Niobium Titan
NIFS	National Institute for Fusion Science
Nb3Sn	Niobiun Tin
PLD	Pulsed laser deposition
PPLP	Polypropylene laminated paper
PPMS	Physical property measurement system
RABiTS	Rolling assisted biaxially textured substrates
RCE	Reactive co-evaporation
*RE*BCO	Rare earth barium copper oxide
s.f.	Self field
SPC	Swiss Plasma Center
SPI	SuperPower Inc.

SST	Shanghai Superconductor Technology
STI	Superconductor Technologies
VDE	Verband der Elektrotechnik Elektronik Informationstechnik e. V.
VESC	Vision Electrics Super Conductors
XLPE	Cross linked polyethylene
YBCO	Yttrium barium copper oxide

C List of symbols

α	Displacement of critical current peaks from 0 °
α	Lay angle
α_i	Coefficient of thermal expansion
a_0 , a_1, a_2	Fit parameter
a_i	Volume fraction
A	Cross section
$\boldsymbol{A}$	Magnetic vector potential
B	Magnetic field
B_c	Critical magnetic field
B_irr	Irreversibility field
B_c1	Meißer phase limit
B_c2	Shubnikov phase limit
B_k	Model parameter within critical current angular magnetic field dependency
c_p	Specific heat capacity
D_h	Hydraulic Diameter
D_outer	Outer diameter of ring type cooling channel
D_inner	Inner diameter of a ring type cooling channel
D_neutral	Diameter of neutral axis
D_CroCo	CroCo Diameter
$D_\mathrm{CroCo,max}$	Maximum CroCo Diameter
d	Helix core diameter
δ	Oxygen contect
δ_{die}	Dieletric thickness
E	Electric field
E_i	Young's modulus

E_{c}	Critical eletric field
E_{BD}	Dieletric breakddown strength
Δh_{vap}	enthalpy of vaporization
Δh	Enthalpy change
I_{c}	Critical current
I_{op}	Operational current
I_{fault}	Fault current
j	Current density
j_{c}	Critrical current density
j_{ce}	Critical engineering current density
j_{c0}	Initial critical current density
k_1	Model parameter within critical current angular magnetic field dependency
k_2	Model parameter within critical current angular magnetic field dependency
k	Thermal conductivity
L	Cable Length
λ	Thermal conductivity
$\dot{m}$	Mass flow rate
n	Fit parametr within the Power Law
n_{fault}	Fault current factor $n_{fault} = \frac{I_{fault}}{I_{op}}$
N	Number of copper cables
θ	Magnetic field angle
ζ	Darcy friction factor
μ	Dynamic viscosity
μ_0	Permeability of vacuum
ρ	Density
ρ_{el}	Specific electric resistance
t_{fault}	Fault duration

ζ	Darcy friction factor
Δp	Pressure drop
ΔT	Temperature rise
Re	Reynolds Number
R_i	Sum of all measured termination resistances
R_{end}	Resistance between cable shoe and copper block
ϵ_ζ	Pipe surface roughness
ε_{100}	Iirreversible strain limit
σ_{100}	Irreversible stress limit
T	Temperature
T_{H}	Ambient Temperature
T_{in}	Coolant inlet temperature
T_{out}	Coolant outlet temperature
T_{c}	Critical temperature
$T_{c,90}$	Critical temperature at 90 % of the nomarlized resistance
$T_{c,10}$	Critical temperature at 10 % of the nomarlized resistance
ΔT_c	Transition width $\Delta T_c = T_{c,90} - T_{c,10}$
T_{start}	Temperature at the beginng of a fault
T_s	Heating temperature
T_{end}	Temperture at the end of a faultw
T_{op}	Opertional temperature
T_{m}	Solidus temperature
p_{in}	Coolant inlet pressure
p_{out}	Coolant outlet pressure
p	Helix lead
q	Heat flow density
q_{cryo}	Heat flow through the cryostat
q_{sc}	Heat load per meter dissipated by the superconductor
Q	Heat flow

R	Electric resistance
r	Radius
ε	Strain
S	Safety factor
σ	stress
v	Coolant velocity
x, y	Cooridnates
U	Voltage

D Publications

Papers

PREUSS, ALAN; WOLF, MICHAEL J.; HEIDUK, MATHIAS; LANGE, CHRISTIAN; FIETZ, WALTER H.: Production and Characterization of Strands for a 35 kA HTS DC Cable Demonstrator. In. *IEEE Transactions on Applied Superconductivity* 29 (2019), Nr. 5, Art. ID. 5402905, DOI 10.1109/TASC.2019.2909208

PREUSS, ALAN; FIETZ, WALTER H.; IMMEL, FABIAN; KAUFFMANN-WEISS, SANDRA; WOLF, MICHAEL J.: Critical Current Degradation of Coated Conductors Under Soldering Conditions. In: *IEEE Transactions on Applied Superconductivity* 28 (2018), Nr. 4, Art. ID. 6601105. DOI 10.1109/TASC.2018.2804094

WOLF, MICHAEL J.;FIETZ, WALTER H.; PREUSS, ALAN: Investigation of HTS CrossConductor Joints, Connectors, and Terminations. In: *IEEE Transactions on Applied Superconductivity* 27 (2017), Nr. 4, Art. ID. 4802605 DOI: 10.1109/TASC.2017.2652305

FIETZ, WALTER H., WOLF, MICHAEL J.; PREUSS, ALAN; HELLER, REINHARD; WEISS, KLAUS-PETER: High-Current HTS Cables: Status and Actual Development. In: *IEEE Transactions on Applied Superconductivity* 26 (2016), Nr. 4, Art. ID: 4800705 DOI 10.1109/TASC.2016.2517319

BAYER, C. M.; GADE, P. V.; BARTH, C.; PREUSS, A.; JUNG, A. WEISS, K. P.: Mechanical reinforcement for RACC cables in high magnetic background fields. In: *Superconductor Science and Technology* 29 (2016), Nr. 2, Art. ID: 025007 DOI 10.1088/0953-2048/29/2/025007

Poster and presentations

PREUSS, ALAN; FIETZ, WALTER H.; IMMEL, FABIAN; WOLF, MICHAEL J.Thermal Stability of commercially available Coated Conductors. 2017 13th European Conference on Applied Superconductivity, Geneva (Switzerland)

PREUSS, ALAN: Conceptual design of high direct current HTS cables. 2017 1st Cambridge – KIT PhD Student Seminar, Cambridge (England)

PREUSS, ALAN; WOLF, MICHAEL J: Application of fusion magnet conductor designs into industrial high current cables. 2017 7th KIT Fusion Ph.D.-Seminar, St. Martin (Germany)

Preuss, Alan; Wolf, Michael J; Fietz, Walter H.: Conceptual design and optimization of HTS DC cables. 2016 Coated Conductors for Applications, Aspen (USA)

Preuss, Alan; Wolf, Michael J; Fietz, Walter H.: Critical current degradation of commercially available coated conductors due to short thermal loads. 2016 Applied Superconductivity Conference, Denver (USA)

Preuss, Alan; Wolf, Michael J; Fietz, Walter H.: High temperature superconducting cables for high current applications. 2016 CEA-TUM Summer School on the Future of a Common European Energy Strategy, Frauenchiemsee (Germany)

E Bibliography

[App+19] K. Appunn et al. "Germany's energy consumption and power mix in charts". In: *Clean Energy Wire* (3.04.2019). URL: https://www.cleanenergywire.org/factsheets/germanys-energy-consumption-and-power-mix-charts.

[Ask18] Askja Energy. "Higher electricity tariffs for smelters in 2017". In: *Askja Energy* (18.01.2018). URL: https://askjaenergy.com/2018/01/18/higher-electricity-tariffs-for-smelters-in-2017/.

[Aug+15] A. Augieri et al. "Electrical Characterization of ENEA High Temperature Superconducting Cable". In: *IEEE TRANSACTIONS ON APPLIED SUPERCONDUCTIVITY* 25.3 (2015), pp. 1–4. DOI: 10.1109/TASC.2014.2364391.

[Bäc17] M. Bäcker. *Cost-effective production of HTS wires by chemical solution deposition.* Geneva, 2017. URL: https://indico.cern.ch/event/659554/contributions/2708209/attachments/1525932/2385971/2MO4-06_Michael_Baecker_Room_34.pdf.

[BBW14] N. Bagrets, C. Barth and K.-P. Weiss. "Low Temperature Thermal and Thermo-Mechanical Properties of Soft Solders for Superconducting Applications". In: *IEEE TRANSACTIONS ON APPLIED SUPERCONDUCTIVITY* 24.3 (2014), pp. 1–3. DOI: 10.1109/TASC.2013.2283869.

[Bal14] H. Balke. *Einführung in die Technische Mechanik: Festigkeitslehre.* 3., aktualisierte Aufl. 2014. SpringerLink : Bücher. Berlin, Heidelberg: Springer Vieweg, 2014. DOI: 10.1007/978-3-642-40981-3.

[BF17] A. Ballarino and R. Flükiger. "Status of MgB2 wire and cable applications in Europe". In: *Journal of Physics: Conference Series* 871.1 (2017), p. 012098. DOI: 10.1088/1742-6596/871/1/012098.

[Bar13] C. Barth. "High Temperature Superconductor Cable Concepts for Fusion Magnets". Ph.D. Thesis. Karlsruhe: Karlsruher Institut für Technologie, 2013.

[BMS15] C. Barth, G. Mondonico and C. Senatore. "Electro-mechanical properties of REBCO coated conductors from various industrial manufacturers at 77K, self-field and 4.2K, 19T". in: *SUPERCONDUCTOR SCIENCE & TECHNOLOGY* 28.4 (2015). DOI: 10.1088/0953-2048/28/4/045011.

[Bar+15] C. Barth et al. "Temperature- and field-dependent characterization of a conductor on round core cable". In: *SUPERCONDUCTOR SCIENCE & TECHNOLOGY* 28.6 (2015). DOI: 10.1088/0953-2048/28/6/065007.

[Bau17] M. Bauer. *Coated conductor HTS tape for energy and magnet applications.* 2017. URL: https://indico.cern.ch/event/659554/contributions/2716305/attachments/1526325/2386699/1MP7-22_Bauer_Markus.pdf.

[Bay+15] C. M. Bayer et al. “Mechanical reinforcement for RACC cables in high magnetic background fields”. In: *Superconductor Science and Technology* 29.2 (2015), p. 025007. DOI: 10.1088/0953-2048/29/2/025007.

[BM86] J. G. Bednorz and K. A. Müller. “Possible highT c superconductivity in the BaLaCuO system”. In: *Z. Physik B - Condensed Matter (Zeitschrift für Physik B Condensed Matter)* 64.2 (1986), pp. 189–193. DOI: 10.1007/BF01303701.

[Bro+] L. Bromberg et al. “CURRENT LEAD OPTIMIZATION FOR CRYOGENIC OPERATION AT INTERMEDIATE TEMPERATURES”. in: (), pp. 577–584. DOI: 10.1063/1.3422405.

[Bru17] C. E. Bruzek. *Status of the MgB2-based high-power DC cable demonstrator within BEST PATHS.* Geneva, 2017. URL: https://indico.cern.ch/event/659554/contributions/2709487/attachments/1527268/2388456/3LO4-06_Christian-Eric_Bruzek_Room_1.pdf.

[BK07] W. Buckel and R. Kleiner. *Supraleitung: Grundlagen und Anwendungen.* 6., vollst. überarb. und erw. Aufl., 1. Nachdr. Lehrbuch Physik. Weinheim: Wiley-VCH, 2007.

[Bul67] F. H. Buller. “Calculation of Electrical Stresses in DC Cable Insulation”. In: *IEEE TRANSACTIONS ON POWER APPARATUS AND SYSTEMS* PAS-86.10 (1967), pp. 1169–1178. DOI: 10.1109/TPAS.1967.291869.

[Bun19] Bundesministerium für Umwelt, Naturschutz, Bau und Reaktorsicherheit. *Klimaschutz in Zahlen: Klimaschutzziele Deutschland und EU.* ed. by Bundesministerium für Umwelt, Naturschutz, Bau und Reaktorsicherheit. 2019. URL: https://www.bmu.de/fileadmin/Daten_BMU/Download_PDF/Klimaschutz/klimaschutz_in_zahlen_klimaziele_bf.pdf.

[Bun] Bundesministerium für Wirtschaft und Technologie. *Energiekonzept für eine umweltschonende, zuverlässige und bezahlbare Energieversorgung.* Ed. by Bundesministerium für Wirtschaft und Technologie. URL: https://www.bmwi.de/Redaktion/DE/Downloads/E/energiekonzept-2010.html.

[Bur19] B. Burger. *Energy Charts.* Ed. by Fraunhofer-Institut für Solare Energiesysteme ISE. 2019. URL: https://www.energy-charts.de/energy_pie_de.htm?year=2018.

[Byk+15] N. Bykovsky et al. “Strain Management in HTS High Current Cables”. In: *IEEE TRANSACTIONS ON APPLIED SUPERCONDUCTIVITY* 25.3 (2015). DOI: 10.1109/TASC.2014.2360041.

[Byk+16] N. Bykovsky et al. “Performance evolution of 60kA HTS cable prototypes in the EDIPO test facility”. In: *SUPERCONDUCTOR SCIENCE & TECHNOLOGY* 29.8 (2016). DOI: 10.1088/0953-2048/29/8/084002.

[Can+92] G. Cannelli et al. “Dynamics of oxygen and phase transitions in the 123 ceramic superconductors by anelastic relaxation measurements”. In: *Superconductor Science and Technology* 5.4 (1992), p. 247. DOI: 10.1088/0953-2048/5/4/009. URL: http://iopscience.iop.org/article/10.1088/0953-2048/5/4/009/pdf.

[Cel+14] G. Celentano et al. “Design of an Industrially Feasible Twisted-Stack HTS Cable-in-Conduit Conductor for Fusion Application”. In: *IEEE TRANSACTIONS ON APPLIED SUPERCONDUCTIVITY* 24.3 (2014). DOI: 10.1109/TASC.2013.2287910.

[Cha+03] J. G. Chandler et al. “Irreversibility fields of Bi-2223 at 30-77 K”. in: *IEEE TRANSACTIONS ON APPLIED SUPERCONDUCTIVITY* 13.2 (2003), pp. 2945–2948. DOI: 10.1109/TASC.2003.812064.

[Cha+] H.-M. Chang et al. “Thermodynamic design of 10 kW Brayton cryocooler for HTS cable”. In: (), pp. 1664–1671. DOI: 10.1063/1.4707099.

[Che+07] B. Chen et al. “Two-dimensional vortices in superconductors”. In: *NATURE PHYSICS* 3.4 (2007), p. 239. DOI: 10.1038/nphys540.

[Che+10] R.-L. Chen et al. “RELIABILITY TEST OF A 1-kW PULSE TUBE CRYOCOOLER FOR HTS CABLE APPLICATION”. in: AIP Conference Proceedings. AIP, 2010, pp. 727–735. DOI: 10.1063/1.3422424.

[Chi+16] N. Chikumoto et al. “Construction and the Circulation Test of the 500-m and 1000-m DC Superconducting Power Cables in Ishikari”. In: *IEEE TRANSACTIONS ON APPLIED SUPERCONDUCTIVITY* 26.3 (2016). DOI: 10.1109/TASC.2016.2537041.

[Cho+11] Y. S. Choi et al. “Progress on the Performance Test of KEPCO HTS Power Cable”. In: *IEEE TRANSACTIONS ON APPLIED SUPERCONDUCTIVITY* 21.3 (2011), pp. 1034–1037. DOI: 10.1109/TASC.2010.2093496.

[Cho+12] Y. S. Choi et al. “THERMAL PROPERTY OF INSULATION MATERIAL FOR HTS POWER CABLE”. in: *ADVANCES IN CRYOGENIC ENGINEERING, VOLS 57A AND 57B* 1434 (2012), pp. 1305–1312. DOI: 10.1063/1.4707055.

[Cli+06] C. C. Clickner et al. "Mechanical properties of pure Ni and Ni-alloy substrate materials for Y-Ba-Cu-O coated superconductors". In: *CRYOGENICS* 46.6 (2006), pp. 432–438. DOI: 10.1016/j.cryogenics.2006.01.014.

[Cry19] Cryomech. *Cryocoolers*. 2019. URL: https://www.cryomech.com/.

[DS13] M. J. Dedicatoria and H. S. Shin. "Analysis on Stress/Strain Tolerances of Ic in Externally Laminated GdBCO CC Tapes". In: *IEEE TRANSACTIONS ON APPLIED SUPERCONDUCTIVITY* 23.3 (2013), p. 8400504. DOI: 10.1109/TASC.2012.2234196.

[Deu] Deutsches Kupferinstitut. *Cu-OFE - Datenblatt*. Ed. by Deutsches Kupferinstitut. URL: https://www.kupferinstitut.de.

[Diz+14] J. R. C. Dizon et al. "Analysis of Thermal Residual Stress/Strain in REBCO Coated Conductor Tapes". In: *IEEE TRANSACTIONS ON APPLIED SUPERCONDUCTIVITY* 24.3 (2014), pp. 1–5. DOI: 10.1109/TASC.2013.2292746.

[DKE07a] DKE Deutsche Kommission Elektrotechnik Elektronik Informationstechnik im DIN und VDE. *Teil 1: Messen des kritischen Stromes - Kritischer Strom (Gleichstrom) von Nb-Ti-Verbundsupraleitern*. 2007. URL: https://www.vde-verlag.de/normen/0390016/din-en-61788-1-vde-0390-1-2007-09.html.

[DKE07b] DKE Deutsche Kommission Elektrotechnik Elektronik Informationstechnik im DIN und VDE. *Teil 2: Messen des kritischen Stromes - Kritischer Strom (Gleichstrom) von Nb_3Sn-Verbundsupraleitern*. 2007. URL: https://www.beuth.de/de/norm/din-en-61788-2/99811113.

[DKE07c] DKE Deutsche Kommission Elektrotechnik Elektronik Informationstechnik im DIN und VDE. *Teil 3: Messen des kritischen Stromes - Kritischer Strom (Gleichstrom) von Ag- und/oder Ag-Legierung ummantelten oxidischen Bi-2212 und Bi-2223-Supraleitern*. 2007. URL: https://www.vde-verlag.de/normen/0390013/din-en-61788-3-vde-0390-3-2007-03.html.

[Dro+15] A. P. Drozdov et al. "Conventional superconductivity at 203 kelvin at high pressures in the sulfur hydride system". In: *Nature* (2015). DOI: 10.1038/nature14964.

[Dro+19] A. P. Drozdov et al. "Superconductivity at 250 K in lanthanum hydride under high pressures". In: *Nature* 569.7757 (2019), p. 528. DOI: 10.1038/s41586-019-1201-8.

[Duv+14] C. Duval et al. "Aluminum smelter comprising electrical conductors made from a superconducting material". US20140209457 A1. 2014.

[Els17] S. Elschner. *Superconducting DC-Busbar for High Current Applications.* Geneva, 2017. URL: https://indico.cern.ch/event/659554/contributions/2709494/attachments/1527269/2388459/3LO4-07_Steffen_Elschner_Room_1.pdf.

[eur19] eurostat. *Preise Elektrizität für Nichthaushaltskunds, ab 2007 - halbjährige Data (nrg_pc_205).* 2019. URL: https://ec.europa.eu/eurostat/data/database?node_code=nrg_pc_205.

[Fes02a] J. E. Fesmire. "Overall thermal performance of flexible piping under simulated bending conditions". In: *AIP Conference Proceedings.* AIP, 2002, pp. 1533–1540. DOI: 10.1063/1.1472187.

[Fes02b] J. E. Fesmire. "Thermal insulation performance of flexible piping for use in HTS power cables". In: *AIP Conference Proceedings.* AIP, 2002, pp. 1525–1532. DOI: 10.1063/1.1472186.

[Fin+08] M. Fink et al. "Measurement of mechanical properties of electronic materials at temperatures down to 4.2 K". in: *CRYOGENICS* 48.11-12 (2008), pp. 497–510. DOI: 10.1016/j.cryogenics.2008.07.006.

[FKG07] B. K. Fitzpatrick, J. T. Kephartl and E. M. Golda. "Characterization of gaseous helium flow cryogen in a flexible cryostat for naval applications of high temperature superconductors". In: *IEEE TRANSACTIONS ON APPLIED SUPERCONDUCTIVITY* 17.2 (2007), pp. 1752–1755. DOI: 10.1109/TASC.2007.897763.

[FM06] H. E. Friedrich and B. L. Mordike, eds. *Magnesium technology: Metallurgy, design data, applications.* Berlin, Heidelberg: Springer, 2006. URL: http://dx.doi.org/10.1007/3-540-30812-1.

[Fuj17] S. Fujita. *Developement of long-length BMO doped REBCO coated conductors by hot-wall PLD process.* Geneva, 2017. URL: https://indico.cern.ch/event/659554/contributions/2708204/attachments/1525930/2385967/2MO4-03_Shinji_Fujita_Room_34.pdf.

[FI02] S. Furuki and G. Ishizuka. "Transformer-Rectifier Package (S-Former) for Aluminum Smelting". In: *Fuji Electric Review* 48.2 (2002). URL: http://www.fujielectric.com/company/tech_archives/pdf/48-02/FER-48-02-045-2002.pdf.

[Gar04] D. E. Garrett. *Handbook of Lithium and Natural Calcium Chloride.* Elsevier Science, 2004. URL: https://books.google.de/books?id=Ua2SVcUBHZgC.

[GE] GE Grid Solutions. *DC Substation Solutions For Metal Processing Applications.* Ed. by General Electric. URL:

http://www.gegridsolutions.com/products/brochures/powerD_vtf/DC_Substation_Solutions_GEA-32019_LR.pdf.

[Gei19] G. A. Geiger. *Die Energieintensiven Industrien in Deutschland.* 2019. URL: http://www.energieintensive.de/startseite.html.

[Ger00] German Institute for Standardization. *Stromschienen aus Kupfer - Bemessung für Dauerstrom.* 1975-12-00. URL: http://perinorm-fr.redi-bw.de/volltexte/CD21DE01/1225654/1225654.pdf?.

[Gol+14] W. Goldacker et al. "Roebel cables from REBCO coated conductors: a one-century-old concept for the superconductivity of the future". In: *SUPERCONDUCTOR SCIENCE & TECHNOLOGY* 27.9 (2014). DOI: 10.1088/0953-2048/27/9/093001.

[Gol04] D. Goloubev. "Kühlung eines resistiven HTSL-Kurzschlussstrombegrenzers mit einer Gemisch-Joule-Thomson-Kältemaschine". Dissertation. Dresden: Technische Universität Dresden, 20.08.2004. URL: http://nbn-resolving.de/urn:nbn:de:swb:14-1095838519812-78347.

[GS12] A. B. Gorospe and H.-S. Shin. "Mechanical Properties of PPLP Material at Cryogenic Temperature". In: *Progress in Superconductivity and Cryogenics* 14.4 (2012), pp. 16–19. DOI: 10.9714/sac.2012.14.4.016.

[GPS04] A. Goyal, M. P. Paranthaman and U. Schoop. "The RABiTS Approach: Using Rolling-Assisted Biaxially Textured Substrates for High-Performance YBCO Superconductors". In: *MRS BULLETIN* 29.08 (2004), pp. 552–561. DOI: 10.1557/mrs2004.161.

[Gra97] P. M. Grant. "Superconductivity and electric power: Promises, promises ... past, present and future". In: *IEEE TRANSACTIONS ON APPLIED SUPERCONDUCTIVITY* 7.2 (1997), pp. 112–132.

[Gre15] M. A. Green. "The cost of coolers for cooling superconducting devices at temperatures at 4.2 K, 20 K, 40 K and 77 K". in: *IOP Conference Series: Materials Science and Engineering* 101.1 (2015), p. 012001. DOI: 10.1088/1757-899X/101/1/012001. URL: https://iopscience.iop.org/article/10.1088/1757-899X/101/1/012001/pdf.

[Gri+14a] F. Grilli et al. "Computation of Losses in HTS Under the Action of Varying Magnetic Fields and Currents". In: *IEEE TRANSACTIONS ON APPLIED SUPERCONDUCTIVITY* 24.1 (2014). DOI: 10.1109/TASC.2013.2259827.

[Gri+14b] F. Grilli et al. "Self-Consistent Modeling of the I-c of HTS Devices: How Accurate do Models Really Need to Be?" In: *IEEE TRANSACTIONS ON APPLIED SUPERCONDUCTIVITY* 24.6 (2014). DOI: 10.1109/TASC.2014.2326925.

[Hah+11] S. Hahn et al. “HTS Pancake Coils Without Turn-to-Turn Insulation”. In: *IEEE TRANSACTIONS ON APPLIED SUPERCONDUCTIVITY* 21.3 (2011), pp. 1592–1595. DOI: 10.1109/TASC.2010.2093492.

[Ham+11] M. Hamabe et al. “Critical Current and Its Magnetic Field Effect Measurement of HTS Tapes Forming DC Superconducting Cable”. In: *IEEE TRANSACTIONS ON APPLIED SUPERCONDUCTIVITY* 21.3 (2011), pp. 1038–1041. DOI: 10.1109/TASC.2010.2089964.

[Hän+07] J. Hänisch et al. “Determination of the irreversibility field of YBCO thin films from pulsed high-field measurements”. In: *Superconductor Science and Technology* 20.3 (2007), p. 228. DOI: 10.1088/0953-2048/20/3/019.

[Haz17] D. Hazelton. *Progress of Wire Development and Process Improvement on 2G HTS at SuperPower.* CICG - Room 3+4 (Geneva), 2017. URL: https://indico.cern.ch/event/659554/contributions/2708203/attachments/1525927/2385960/2MO4-02_Drew_Hazelton_Room_34.pdf.

[Haz+02] M. Hazeyama et al. “Partial discharge inception characteristics under butt gap condition in liquid nitrogen/PPLP (R) composite insulation system for high temperature superconducting cable”. In: *IEEE TRANSACTIONS ON DIELECTRICS AND ELECTRICAL INSULATION* 9.6 (2002), pp. 939–944.

[HKI14] F. Herzog, T. Kutz and Int Inst Refrigerat. “Cooling Unit for Superconducting Electricity Cables”. In: *13TH CRYOGENICS 2014 IIR INTERNATIONAL CONFERENCE* 2014.1 (2014), pp. 99–104.

[Her18] F. Herzog. *Betrieb und Zuverlässigkeit von Flüssigstickstoff-Kühlsystemen für supraleitende Kabel.* Berlin, 21.03.2018. URL: https://ivsupra.de/wp-content/uploads/2019/01/9-ZIEHL-herzog.pdf.

[Her+16] F. Herzog et al. “Cooling unit for the AmpaCity project – One year successful operation”. In: *CRYOGENICS* 80 (2016), pp. 204–209. DOI: 10.1016/j.cryogenics.2016.04.001.

[Hol+13] M. Hollik et al. “Design of electronic measurement and quench detection equipment for the Current Lead Test facility Karlsruhe (CuLTKa)”. In: *FUSION ENGINEERING AND DESIGN* 88.6-8 (2013), pp. 1445–1448. DOI: 10.1016/j.fusengdes.2013.02.082.

[Hol17] B. Holzapfel. *Status of Industrial Coated Conductor Production and Properties.* Geneva, 2017. URL: https://indico.cern.ch/event/659554/contributions/2708085/attachments/1525495/2387421/2MO1-02_EUCAS_Geneva_2017_Holzapfel_pub.pdf.

[HYL12] K.-J. Hong, C. Yi and Y.-k. Lee. "Geometry and friction of helically wrapped wires in a cable subjected to tension and bending". In: *INTERNATIONAL JOURNAL OF STEEL STRUCTURES* 12.2 (2012), pp. 233–242. DOI: 10.1007/s13296-012-2007-9.

[Hwa+13] J.-S. Hwang et al. "Evaluation of DC electric field distribution of PPLP specimen based on the measurement of electrical conductivity in LN2". In: *PHYSICA C-SUPERCONDUCTIVITY AND ITS APPLICATIONS* 494 (2013), pp. 307–310. DOI: 10.1016/j.physc.2013.04.005.

[IEZ07] K. Inagaki, J. Ekh and S. Zahrai. "Mechanical analysis of second order helical structure in electrical cable". In: *INTERNATIONAL JOURNAL OF SOLIDS AND STRUCTURES* 44.5 (2007), pp. 1657–1679. DOI: 10.1016/j.ijsolstr.2006.06.045.

[Ito+14] S. Ito et al. "Performance of a Mechanical Bridge Joint for 30-kA-Class High-Temperature Superconducting Conductors". In: *IEEE TRANSACTIONS ON APPLIED SUPERCONDUCTIVITY* 24.3 (2014). DOI: 10.1109/TASC.2013.2291157.

[Iva+16] Y. Ivanov et al. "Current Imbalance and AC Losses of Long-Distance DC HTS Cable". In: *IEEE TRANSACTIONS ON APPLIED SUPERCONDUCTIVITY* 26.7 (2016).

[JM98] M. J. P. Jeroense and P. H. F. Morshuis. "Electric fields in HVDC paper-insulated cables". In: *IEEE TRANSACTIONS ON DIELECTRICS AND ELECTRICAL INSULATION* 5.2 (1998), pp. 225–236. DOI: 10.1109/94.671940.

[Ji+15] W.-T. Ji et al. "Summary and evaluation on single-phase heat transfer enhancement techniques of liquid laminar and turbulent pipe flow". In: *INTERNATIONAL JOURNAL OF HEAT AND MASS TRANSFER* 88 (2015), pp. 735–754. DOI: 10.1016/j.ijheatmasstransfer.2015.04.008.

[Jia+11] J. Jiang et al. "Doubled critical current density in Bi-2212 round wires by reduction of the residual bubble density". In: *Superconductor Science and Technology* 24.8 (2011), p. 082001. DOI: 10.1088/0953-2048/24/8/082001.

[Kab] Kabeltrommel GmbH & Co. KG. *Produktinformation Standard Holzspule Typ 281.* URL: http://www.kabeltrommel.de/uploads/media/Typ_281.pdf.

[Kam11] H. Kamerlingh Onnes. "Further experiments with Liquid Helium G. On the electrical resistance of Pure Metals etc. VI. On the Sudden Change in the Rate at which the Resistance of Mercury Disappears". In: *Koninklijke Nederlandse Akademie van Wetenschappen Proceedings Series B Physical Sciences* 14 (1911), pp. 818–821.

[Kam+06] Y. Kamihara et al. "Iron-based layered superconductor: LaOFeP". in: *JOURNAL OF THE AMERICAN CHEMICAL SOCIETY* 128.31 (2006), pp. 10012–10013. DOI: 10.1021/ja063355c.

[KC17] S. Kasap and P. Capper. *Springer Handbook of Electronic and Photonic Materials.* Cham: Springer International Publishing, 2017. DOI: 10.1007/978-3-319-48933-9.

[Kau71] K. Kauder. "Strömungs- und Widerstandsverhalten in gewellten Rohren". Dissertation. Hannover: Technische Universität Hannover, 1971.

[Kep+11] J. T. Kephart et al. "High Temperature Superconducting Degaussing From Feasibility Study to Fleet Adoption". In: *IEEE TRANSACTIONS ON APPLIED SUPERCONDUCTIVITY* 21.3 (2011), pp. 2229–2232. DOI: 10.1109/TASC.2010.2092746.

[Kik+15] Y. Kikuchi et al. "Partial Discharge Characteristics in Composite Insulation Systems with PPLP (R) for HTS Cable". In: *IEEE TRANSACTIONS ON DIELECTRICS AND ELECTRICAL INSULATION* 22.2 (2015), pp. 1025–1030. DOI: 10.1109/TDEI.2014.004657.

[Kim+08] M. J. Kim et al. "Determination of maximum permissible temperature rise considering repetitive over-current characteristics of YBCO coated conductors". In: *IEEE TRANSACTIONS ON APPLIED SUPERCONDUCTIVITY* 18.2 (2008), pp. 660–663. DOI: 10.1109/TASC.2008.921397.

[Kim+12] S. H. Kim et al. "Electrical Insulation Characteristics of PPLP as a HTS DC Cable Dielectric and GFRP as Insulating Material for Terminations". In: *IEEE TRANSACTIONS ON APPLIED SUPERCONDUCTIVITY* 22.3 (2012). DOI: 10.1109/TASC.2011.2181470.

[Kim+13a] S. H. Kim et al. "DC and Impulse Insulation Characteristics of PPLP for HTS DC Cable". In: *Journal of the Korean Institute of Electrical and Electronic Material Engineers* 26.7 (2013), pp. 545–549. DOI: 10.4313/JKEM.2013.26.7.545.

[Kim+13b] S.-K. Kim et al. "Design and AC Loss Analysis of a 22.9 kV/50 MVA Class Triaxial HTS Power Cable". In: *JOURNAL OF SUPERCONDUCTIVITY AND NOVEL MAGNETISM* 26.4 (2013), pp. 755–758. DOI: 10.1007/s10948-012-1868-7.

[Kim+13c] S.-K. Kim et al. "Development and Performance Analysis of a 22.9 kV/50 MVA Tri-axial HTS Power Cable Core". In: *IEEE TRANSACTIONS ON APPLIED SUPERCONDUCTIVITY* 23.3 (2013). DOI: 10.1109/TASC.2013.2238286.

[Kim+15] W.-J. Kim et al. "Insulation Characteristics of PPLP and Design of 250 kV Class HTS DC Cable". In: *IEEE TRANSACTIONS ON APPLIED SUPERCONDUCTIVITY* 25.3 (2015). DOI: 10.1109/TASC.2015.2389150.

[KGO15] T. M. Kochenburger, S. Grohmann and L. R. Oellrich. "Evaluation of a Two-stage Mixed Refrigerant Cascade for HTS Cooling Below 60 K". in: *Physics Procedia* 67 (2015), pp. 227–232. DOI: 10.1016/j.phpro.2015.06.039.

[Kos+12] V. V. Kostyuk et al. "Experimental hybrid power transmission line with liquid hydrogen and MgB2-based superconducting cable". In: *TECHNICAL PHYSICS LETTERS* 38.3 (2012), pp. 279–282. DOI: 10.1134/S106378501203025X.

[Kos+15] V. V. Kostyuk et al. "Cryogenic design and test results of 30-m flexible hybrid energy transfer line with liquid hydrogen and superconducting MgB2 cable". In: *CRYOGENICS* 66 (2015), pp. 34–42. DOI: 10.1016/j.cryogenics.2014.11.010.

[Kov+15] O. A. Kovalchuk et al. "Cryogenic High Voltage Insulation Breaks for ITER". in: *Physics Procedia* 67 (2015), pp. 698–704. DOI: 10.1016/j.phpro.2015.06.118.

[KD14] H. Kvande and P. A. Drabløs. "The aluminum smelting process and innovative alternative technologies". In: *Journal of occupational and environmental medicine / American College of Occupational and Environmental Medicine* 56.5 Suppl (2014), S23–32. DOI: 10.1097/JOM.0000000000000062.

[Lar+01] D. Larbalestier et al. "High-Tc superconducting materials for electric power applications". In: *Nature* 414.6861 (2001), pp. 368–377. DOI: 10.1038/35104654.

[Lee+13] B.-W. Lee et al. "Comparison Between PD Inception Voltage and BD Voltage of PPLP in LN2 Considering HTS Cable Insulation". In: *IEEE TRANSACTIONS ON APPLIED SUPERCONDUCTIVITY* 23.3 (2013). DOI: 10.1109/TASC.2013.2240031.

[Lee17] H. Lee. *Progress in SuNAM's Coated Conductor Manufacturing*. Geneva, 2017. URL: https://indico.cern.ch/event/659554/contributions/2708208/attachments/1525931/2385968/2MO4-05_Hunju_Lee_Room_34.pdf.

[Lee+11] S. Lee et al. "Modeling of a 22.9 kV 50 MVA superconducting power cable based on PSCAD/EMTDC for application to the Icheon substation in Korea: The 23rd International Symposium on Superconductivity". In: *Physica C: Superconductivity* 471.21–22 (2011), pp. 1283–1289. DOI: 10.1016/j.physc.2011.05.179.

[Li+13] Y. X. Li et al. "Modeling for mechanical response of CICC by hierarchical approach and ABAQUS simulation". In: *FUSION ENGINEERING AND DESIGN* 88.11 (2013), pp. 2907–2917. DOI: 10.1016/j.fusengdes.2013.06.002.

[Lin19] Linde AG. *Cryogenic Standard Tanks*. 2019. URL: www.linde-engineering.com.

[Lin97] P. Linstrom. *NIST Chemistry WebBook, NIST Standard Reference Database 69.* Gaithersburg, 1997. DOI: 10.18434/T4D303.

[LCZ08] J. Lu, E. S. Choi and H. D. Zhou. "Physical properties of Hastelloy (R) C-276 (TM) at cryogenic temperatures". In: *JOURNAL OF APPLIED PHYSICS* 103.6 (2008). DOI: 10.1063/1.2899058.

[Lu+11] J. Lu et al. "Lap Joint Resistance of YBCO Coated Conductors". In: *IEEE TRANSACTIONS ON APPLIED SUPERCONDUCTIVITY* 21.3 (2011), pp. 3009–3012. DOI: 10.1109/TASC.2010.2091934.

[Lub83] M. Lubell. "Empirical scaling formulas for critical current and critical field for commercial NbTi". In: *IEEE TRANSACTIONS ON MAGNETICS* 19.3 (1983), pp. 754–757. DOI: 10.1109/TMAG.1983.1062311.

[Mag+11] J. F. Maguire et al. "Progress and Status of a 2G HTS Power Cable to Be Installed in the Long Island Power Authority (LIPA) Grid". In: *IEEE TRANSACTIONS ON APPLIED SUPERCONDUCTIVITY* 21.3 (2011), pp. 961–966. DOI: 10.1109/TASC.2010.2093108.

[MYR15] A. P. Malozemoff, J. Yuan and C. M. Rey. "High-temperature superconducting (HTS) AC cables for power grid applications". In: *Superconductors in the Power Grid: Materials and Applications.* Ed. by C. M. Rey. 2015, pp. 133–188. DOI: 10.1016/B978-1-78242-029-3.00005-4.

[Man11] A. Mann. "High-temperature superconductivity at 25: Still in suspense". In: *Nature* 475.7356 (2011), pp. 280–282. DOI: 10.1038/475280a.

[Mas+05] T. Masuda et al. "Design and experimental results for Albany HTS cable". In: *IEEE TRANSACTIONS ON APPLIED SUPERCONDUCTIVITY* 15.2 (2005), pp. 1806–1809. DOI: 10.1109/TASC.2005.849296.

[Mat+09] V. Matias et al. "Reactive Co-Evaporation of YBCO as a Low-Cost Process for Fabricating Coated Conductors". In: *IEEE TRANSACTIONS ON APPLIED SUPERCONDUCTIVITY* 19.3 (2009), pp. 3172–3175. DOI: 10.1109/TASC.2009.2019212.

[McF59] R. McFee. "Optimum Input Leads for Cryogenic Apparatus". In: *REVIEW OF SCIENTIFIC INSTRUMENTS* 30.2 (1959), pp. 98–102. DOI: 10.1063/1.1716499.

[Met] MetOx. *Revolutionary Technology – MetOx.* URL: http://www.metoxtech.com/revolutionary-technology/.

[Mia+05] H. Miao et al. "Development of Round Multifilament Bi-2212/Ag Wires for High Field Magnet Applications". In: *IEEE TRANSACTIONS ON APPLIED*

SUPERCONDUCTIVITY 15.2 (2005), pp. 2554–2557. DOI: 10.1109/TASC.2005.847648.

[Mic+15] P. C. Michael et al. "Design and test of a prototype 20 kA HTS DC power transmission cable". In: *IEEE TRANSACTIONS ON APPLIED SUPERCONDUCTIVITY* 25.3 (2015). DOI: 10.1109/TASC.2014.2373824.

[Min+09] J. V. Minervini et al. *Superconducting DC PowerTransmission and Distribution: Final Report to the MIT Energy Council.* Ed. by Plasma Science and Fusion Center. 2009.

[Mit05] N. Mitchell. "Finite element simulations of elasto-plastic processes in Nb3Sn strands". In: *CRYOGENICS* 45.7 (2005), pp. 501–515. DOI: 10.1016/j.cryogenics.2005.06.003.

[Mol17] A. Molodyk. *Development of 2G HTS wire production at SuperOx.* Geneva, 2017. URL: https://indico.cern.ch/event/659554/contributions/2708207/attachments/1525929/2385959/2MO4-04_Alexander_Molodyk_Room_34.pdf.

[Mor15] A. Morandi. "HTS dc transmission and distribution: concepts, applications and benefits". In: *Superconductor Science and Technology* 28.12 (2015), p. 123001. DOI: 10.1088/0953-2048/28/12/123001.

[Muk+09] S. Mukoyama et al. "Development of YBCO High-Tc Superconducting Power Cables". In: *Furukawa Review* 35 (2009), pp. 18–22. URL: https://www.furukawa.co.jp/review/fr035/fr35_04.pdf.

[Mul+15] T. Mulder et al. "Optimized and practical electrical joints for CORC type HTS cables". In: *IOP Conference Series: Materials Science and Engineering* 102.1 (2015), p. 012026. DOI: 10.1088/1757-899X/102/1/012026.

[Mul+17] T. Mulder et al. "Demonstration of the ReBCO CORC 47kA@10T/4.2K Cable-In-Conduit-Conductor and its Joint Terminals at 4.5 and 77 K". in: *IEEE TRANSACTIONS ON APPLIED SUPERCONDUCTIVITY* 27.4 (2017). DOI: 10.1109/TASC.2017.2652300.

[Mul+16a] T. Mulder et al. "Design and Manufacturing of a 45 kA at 10 T REBCO-CORC Cable-in-Conduit Conductor for Large-Scale Magnets". In: *IEEE TRANSACTIONS ON APPLIED SUPERCONDUCTIVITY* 26.4 (2016). DOI: 10.1109/TASC.2016.2527241.

[Mul+16b] T. Mulder et al. "Development of Joint Terminals for a New Six-Around-One ReBCO-CORC Cable-in-Conduit Conductor Rated 45 kA at 10 T/4 K". in: *IEEE TRANSACTIONS ON APPLIED SUPERCONDUCTIVITY* 26.3 (2016). DOI: 10.1109/TASC.2016.2530875.

[Mul+16c] T. Mulder et al. "Performance Test of an 8 kA @ 10-T 4.2-K ReBCO-CORC Cable". In: *IEEE TRANSACTIONS ON APPLIED SUPERCONDUCTIVITY* 26.4 (2016). DOI: 10.1109/TASC.2016.2545115.

[Mur+18] J. S. Murtomaki et al. "10 kA Joints for HTS Roebel Cables". In: *IEEE TRANSACTIONS ON APPLIED SUPERCONDUCTIVITY* 28.3 (2018), pp. 1–6. DOI: 10.1109/TASC.2018.2804951.

[Nag+01] J. Nagamatsu et al. "Superconductivity at 39 K in magnesium diboride". In: *Nature* 410.6824 (2001), pp. 63–64. DOI: 10.1038/35065039.

[Nak+98] H. Nakagawa et al. "DC and AC magneto-resistance measurement technique for YBa2Cu3O7−δ thin films in megagauss fields". In: *Physica B: Condensed Matter* 246-247 (1998), pp. 429–432. DOI: 10.1016/S0921-4526(97)00953-8.

[Neu04] H. Neumann. "Concept for thermal insulation arrangement within a flexible cryostat for HTS power cables". In: *CRYOGENICS* 44.2 (2004), pp. 93–99. DOI: 10.1016/j.cryogenics.2003.08.005.

[Nex06] Nexans. *Niederspannungsnetzkabel und Mittelspannungskabel.* Ed. by Nexans Schweiz AG. 2006. URL: http://www.nexans.com/Switzerland/files/NEXANS06_BTMTAcc_D.pdf.

[Nex] Nexans. *CRYOFLEX® Transferleitungssysteme für LN2.* Ed. by Nexans. URL: http://www.nexans.de/eservice/Germany-de_DE/navigateproduct_540368851/CRYOFLEX_Transfer_Lines_for_LN2.html#description.

[NOR70] W. T. NORRIS. "CALCULATION OF HYSTERESIS LOSSES IN HARD SUPERCONDUCTORS CARRYING AC - ISOLATED CONDUCTORS AND EDGES OF THIN SHEETS". in: *JOURNAL OF PHYSICS D-APPLIED PHYSICS* 3.4 (1970), pp. 489–&.

[OBH07] T. F. O'Brien, T. V. Bommaraju and F. Hine. *Handbook of Chlor-Alkali Technology.* Developments in Hydrobiology. Dordrecht: Springer, 2007. URL: http://gbv.eblib.com/patron/FullRecord.aspx?p=603758.

[Oku+02] H. Okubo et al. "V-t characteristics of partial discharge inception in liquid nitrogen/PPLP (R) composite insulation system for HTS cable". In: *IEEE TRANSACTIONS ON DIELECTRICS AND ELECTRICAL INSULATION* 9.6 (2002), pp. 945–951.

[Ord+13] J. C. Ordonez et al. "Temperature and Pressure Drop Model for Gaseous Helium Cooled Superconducting DC Cables". In: *IEEE TRANSACTIONS ON APPLIED SUPERCONDUCTIVITY* 23.3 (2013). DOI: 10.1109/TASC.2013.2241380.

[OMH16] K. Osamura, S. Machiya and D. P. Hampshire. "Mechanism for the uniaxial strain dependence of the critical current in practical REBCO tapes". In: *Superconductor Science and Technology* 29.6 (2016), p. 065019. DOI: 10.1088/0953-2048/29/6/065019.

[Ott+15] S. Otten et al. "Enhancement of the transverse stress tolerance of REBCO Roebel cables by epoxy impregnation". In: *SUPERCONDUCTOR SCIENCE & TECHNOLOGY* 28.6 (2015). DOI: 10.1088/0953-2048/28/6/065014.

[Ott+16] S. Otten et al. "Bending properties of different REBCO coated conductor tapes and Roebel cables at T = 77 K". in: *Superconductor Science and Technology* 29.12 (2016), p. 125003. DOI: 10.1088/0953-2048/29/12/125003.

[Pap97] K. O. Papailiou. "On the bending stiffness of transmission line conductors". In: *IEEE TRANSACTIONS ON POWER DELIVERY* 12.4 (1997), pp. 1576–1588.

[PI04] M. P. Paranthaman and T. Izumi. "High-performance YBCO-coated superconductor wires". In: *MRS BULLETIN* 29.8 (2004), pp. 533–536.

[Par+11] E. Pardo et al. "Low-magnetic-field dependence and anisotropy of the critical current density in coated conductors". In: *SUPERCONDUCTOR SCIENCE & TECHNOLOGY* 24.6 (2011). DOI: 10.1088/0953-2048/24/6/065007.

[Par+14] Y. Park et al. "A superconducting joint for GdBa2Cu3O7−δ-coated conductors". In: *NPG Asia Materials* 6.5 (2014), e98. DOI: 10.1038/am.2014.18. URL: https://www.nature.com/articles/am201418.pdf.

[PKA07] A. Pratt, P. Kumar and T. V. Aldridge. *Evaluation of 400V DC distribution in telco and data centers to improve energy efficiency.* IEEE, 2007. DOI: 10.1109/INTLEC.2007.4448733.

[Pre+18] A. Preuss et al. "Critical Current Degradation of Coated Conductors Under Soldering Conditions". In: *IEEE TRANSACTIONS ON APPLIED SUPERCONDUCTIVITY* 28.4 (2018), pp. 1–5. DOI: 10.1109/TASC.2018.2804094.

[Pre+19] A. Preuss et al. "Production and Characterization of Strands for a 35 kA HTS DC Cable Demonstrator". In: *IEEE TRANSACTIONS ON APPLIED SUPERCONDUCTIVITY* 29.5 (2019), pp. 1–5. DOI: 10.1109/TASC.2019.2909208.

[Rad09] R. Radebaugh. "Cryocoolers: the state of the art and recent developments". In: *JOURNAL OF PHYSICS-CONDENSED MATTER* 21.16 (2009). DOI: 10.1088/0953-8984/21/16/164219.

[Rei14] W. Reiser. *Hochtemperatur-Supraleitung für industrielle Hochstromschienen.* 2014.

[Ren+08] Z.-A. Ren et al. "Superconductivity and phase diagram in iron-based arsenic-oxides ReFeAsO1−δ (Re = rare-earth metal) without fluorine doping". In: *EPL (Europhysics Letters)* 83.1 (2008), p. 17002. DOI: 10.1209/0295-5075/83/17002.

[Ren+16] S. Renaudier et al. "Aluminium smelter comprising a compensating electric circuit". EP3030695B1. 2016.

[Rey15] C. M. Rey, ed. *Superconductors in the Power Grid: Materials and Applications.* 2015.

[Rez+10] F. Rezaeifar et al. "Characterization of Partial Discharge in Composite Insulation System with PPLP (R) for HTS Cable". In: *IEEE TRANSACTIONS ON DIELECTRICS AND ELECTRICAL INSULATION* 17.6 (2010), pp. 1747–1753.

[RRM02] E. Romeo, C. Royo and A. Monzon. "Improved explicit equations for estimation of the friction factor in rough and smooth pipes". In: *CHEMICAL ENGINEERING JOURNAL* 86.3 (2002), pp. 369–374.

[RUN95] M. RUNDE. "APPLICATION OF HIGH-T-C SUPERCONDUCTORS IN ALUMINUM ELECTROLYSIS PLANTS". in: *IEEE TRANSACTIONS ON APPLIED SUPERCONDUCTIVITY* 5.2 (1995), pp. 813–816.

[Sam] Samri Advanced Materials. *Company Profile - Handout.* Ed. by Samri Advanced Materials. 2018 Applied Superconductivity Conference.

[SWB15] J. Sas, K. P. Weiss and N. Bagrets. "CRYOMAK – THE OVERVIEW OF CRYOGENIC TESTING FACILITIES IN KARLSRUHE". in: *Acta Metallurgica Slovaca* 21.4 (2015), p. 330. DOI: 10.12776/ams.v21i4.650.

[Sau17] J.-M. Saugrain. *Three years operation experience of the AmpaCity system installation in Essen, Germany.* Geneva, 2017. URL: https://indico.cern.ch/event/659554/contributions/2709480/attachments/1527270/2388461/3LO4-05_Jean-Maxime_Saugrain_Room_1.pdf.

[Sch17] S. I. Schlachter. *Design Aspects for DC-HTS Cables in Hybrid-Electric Propulsion Systems for Aircraft.* Madison, USA, 11.07.2017. URL: https://indico.cern.ch/event/578092/contributions/2537822/attachments/1491653/2345743/M2OrE-04_Schlachter.pdf.

[Sch+12] F. Schmidt et al. "Operation Experience and further Development of a High-Temperature Superconducting Power Cable in the Long Island Power Authority Grid". In: *SUPERCONDUCTIVITY CENTENNIAL CONFERENCE 2011* 36 (2012), pp. 1137–1144. DOI: 10.1016/j.phpro.2012.06.190.

[Sch+17] F. Schreiner et al. "Design and Manufacturing of a Multistage Cooled Current Lead for Superconducting High Current DC Busbars in Industrial Applications". In:

IEEE TRANSACTIONS ON APPLIED SUPERCONDUCTIVITY 27.4 (2017), pp. 1–5. DOI: 10.1109/TASC.2017.2655108.

[Sei15] P. Seidel. *Applied Superconductivity: Handbook on devices and applications.* Weinheim, Germany: Wiley-VCH Verlag GmbH & Co. KGaA, 2015. DOI: 10.1002/9783527670635.

[Sha+17] E. Shabagin et al. "Modelling of 3D temperature profiles and pressure drop in concentric three-phase HTS power cables". In: *CRYOGENICS* 81 (2017), pp. 24–32. DOI: 10.1016/j.cryogenics.2016.11.004.

[Sha15] R. G. Sharma. *Superconductivity: Basics and Applications to Magnets.* Vol. 214. Cham: Springer International Publishing, 2015. DOI: 10.1007/978-3-319-13713-1.

[SHI19] SHI Cryogenics Group. *Cryocoolers.* 2019. URL: http://www.shicryogenics.com/products/specialty-cryocoolers/ch-110-77k-cryocooler-series/.

[Shi17] Y. Shirai. *Prospect of Liquid Hydrogen Cooled Superconducting Power Apparatus and Carbon Free Energy System.* Geneva, 2017. URL: https://indico.cern.ch/event/659554/contributions/2709415/attachments/1527135/2388190/3LO1-02_Yasuyuki_Shirai_Room_1.pdf.

[Sim+13] K. Sim et al. "Design and Current Transporting Characteristics of 80 kV Direct Current High Temperature Superconducting Cable Core". In: *IEEE TRANSACTIONS ON APPLIED SUPERCONDUCTIVITY* 23.3 (2013). DOI: 10.1109/TASC.2012.2237001.

[SDR92] N. J. Simon, E. S. Drexler and R. P. Reed. *Properties of Copper and Copper Alloys at Cryogenic Temperatures: NIST Monograph 177.* Boulder, 1992. URL: http://nvlpubs.nist.gov/nistpubs/Legacy/MONO/nistmonograph177.pdf.

[ST04] U. K. Sinha and J. Tolbert. "Superconducting cable termination". US6936771B2. 2004.

[Spa00] R. Span. "A Reference Equation of State for the Thermodynamic Properties of Nitrogen for Temperatures from 63.151 to 1000 K and Pressures to 2200 MPa". In: *Journal of Physical and Chemical Reference Data* 29.6 (2000), p. 1361. DOI: 10.1063/1.1349047.

[Ste15] M. Stemmle. *Superconducting Power Cables.* Karlsruhe, 22.07.2015.

[Ste+13] M. Stemmle et al. "AmpaCity — Installation of advanced superconducting 10 kV system in city center replaces conventional 110 kV cables". In: *Proceedings of 2013*

IEEE International Conference on Applied Superconductivity and Electromagnetic Devices (ASEMD) (2013), pp. 323–326. DOI: 10.1109/ASEMD.2013.6780785.

[Sti19] Stirling Cryogenics. *SPC-4 Cryogenerator.* 2019. URL: https://www.stirling%20cryogenics.eu/en/products/cryocoolers/spc-4-cryogenerator.

[Str] T. R. Strobridge. *Cryogenic refrigerators: An updated survey.* Ed. by National Institute of Standards and Technology. URL: https://www.gpo.gov/fdsys/pkg/GOVPUB-C13-6e10d382548a9b157a6ccac39043f47c/pdf/GOVPUB-C13-6e10d382548a9b157a6ccac39043f47c.pdf.

[SN16] Stuart Wimbush and Nick Strickland. *A high-temperature superconducting (HTS) wire critical current database.* 2016. DOI: 10.6084/M9.FIGSHARE.C.2861821.V2.

[Sun+13] J. Sun et al. "Critical current behavior of a BSCCO tape in the stacked conductors under different current feeding mode". In: *PHYSICA C-SUPERCONDUCTIVITY AND ITS APPLICATIONS* 494 (2013), pp. 297–301. DOI: 10.1016/j.physc.2013.04.031.

[Syt+15] V. E. Sytnikov et al. "Status of HTS cable link project for St. Petersburg Grid". In: *IEEE TRANSACTIONS ON APPLIED SUPERCONDUCTIVITY* 25.3 (2015). DOI: 10.1109/TASC.2014.2373814.

[Tak+08] H. Takahashi et al. "Superconductivity at 43 K in an iron-based layered compound LaO1-xFxFeAs". In: *Nature* 453.7193 (2008), p. 376.

[TCM14] M. Takayasu, L. Chiesa and J. V. Minervini. "Development of Termination Methods for 2G HTS Tape Cable Conductors". In: *IEEE TRANSACTIONS ON APPLIED SUPERCONDUCTIVITY* 24.3 (2014). DOI: 10.1109/TASC.2013.2280839.

[Tak+12] M. Takayasu et al. "INVESTIGATION OF TWISTED STACKED-TAPE CABLE CONDUCTOR". in: *ADVANCES IN CRYOGENIC ENGINEERING, VOL 58* 1435 (2012), pp. 273–280. DOI: 10.1063/1.4712106.

[Tak+13] M. Takayasu et al. "Conductor Characterization of YBCO Twisted Stacked-Tape Cables". In: *IEEE TRANSACTIONS ON APPLIED SUPERCONDUCTIVITY* 23.3 (2013). DOI: 10.1109/TASC.2012.2234182.

[Tak+17] M. Takayasu et al. "Investigation of HTS Twisted Stacked-Tape Cable (TSTC) Conductor for High-Field, High-Current Fusion Magnets". In: *IEEE TRANSACTIONS ON APPLIED SUPERCONDUCTIVITY* 27.4 (2017), pp. 1–5. DOI: 10.1109/TASC.2017.2652328.

[Ted+12] R. Tediosi et al. "Low Temperature and Magnetic Field Performance of Spliced Commercial YBCO CC". in: *IEEE TRANSACTIONS ON APPLIED*

SUPERCONDUCTIVITY 22.3 (2012), p. 6600804. DOI: 10.1109/TASC.2011.2178579.

[Ter+14] Y. Terazaki et al. "Critical Current Measurement of 30 kA-Class HTS Conductor Samples". In: *IEEE TRANSACTIONS ON APPLIED SUPERCONDUCTIVITY* 24.3 (2014). DOI: 10.1109/TASC.2013.2287715.

[Ter+15] Y. Terazaki et al. "Measurement and Analysis of Critical Current of 100-kA Class Simply-Stacked HTS Conductors". In: *IEEE TRANSACTIONS ON APPLIED SUPERCONDUCTIVITY* 25.3 (2015). DOI: 10.1109/TASC.2014.2377793.

[Ter17] Y. Terazaki. *Stability Analysis of the 100 kA-class HTS STARS Conductor.* Geneva, 2017. URL: https://indico.cern.ch/event/659554/contributions/2708136/attachments/1525769/2385680/2LO3-05_Yoshiro_Terazaki_Room_1.pdf.

[TBE19] U. Terstege, M. Bitz and J. Ewert. *Investitionsrechnung klipp & klar.* Wiesbaden: Springer Fachmedien Wiesbaden, 2019. DOI: 10.1007/978-3-658-20992-6.

[The19] The London Metal Exchange. *LME Aluminium - Average Prices.* 2019. URL: https://www.lme.com/.

[Tis+99] V. G. Tissen et al. "Pressure dependence of Tc in strongly underdoped YBa2Cu3O6.41 as a function of pressure–temperature history". In: *Physica C: Superconductivity* 316.1-2 (1999), pp. 21–26. DOI: 10.1016/S0921-4534(99)00261-0.

[Tom+12] M. Tomita et al. "Development of 10 kA high temperature superconducting power cable for railway systems". In: *JOURNAL OF APPLIED PHYSICS* 111.6 (2012). DOI: 10.1063/1.3696975.

[Tre06] L. Trevisani. "Design and simulation of a large scale energy storage and power transmission system for remote renewable energy sources exploitation". Doctoral Thesis. Bologna: University of Bologna, 3.2006. URL: http://www.die.ing.unibo.it/dottorato_it/Trevisani/Trevisani_PhD-thesis.pdf.

[Tri18] Trimet Aluminium SE. *Geschäftsbericht 2016/17.* Ed. by Trimet Aluminium SE. Essen, 2018. URL: https://www.trimet.eu/de/geschaeftsberichte/trimet-geschaeftsbericht-2017.pdf.

[Tri19] Trimet Aluminium SE. *Geschäftsbericht 2017/18.* Ed. by Trimet Aluminium SE. Essen, 2019. URL: https://www.trimet.eu/de/geschaeftsberichte/trimet-geschaeftsbericht-2018.pdf.

[Tsu+17] K. Tsuchiya et al. “Critical current measurement of commercial REBCO conductors at 4.2 K”. in: *CRYOGENICS* 85 (2017), pp. 1–7. DOI: 10.1016/j.cryogenics.2017.05.002.

[Ugl+15] D. Uglietti et al. “Test of 60kA coated conductor cable prototypes for fusion magnets”. In: *SUPERCONDUCTOR SCIENCE & TECHNOLOGY* 28.12 (2015). DOI: 10.1088/0953-2048/28/12/124005.

[Uso+18] A. Usoskin et al. “Double-Disordered HTS-Coated Conductors and Their Assemblies Aimed for Ultra-High Fields: Large Area Tapes”. In: *IEEE TRANSACTIONS ON APPLIED SUPERCONDUCTIVITY* 28.4 (2018), pp. 1–6. DOI: 10.1109/TASC.2018.2801348.

[van14] van der Laan, D. C. *Conductor on Round Core (CORC) cable development for accelerator magnets.* Hamburg, Germany, 22.05.2014. URL: http://indico.cern.ch/event/308828/session/3/contribution/15/attachments/589812/811817/van_der_Laan_WAMHTS_2014_Final.pdf.

[vE07] van der Laan, D. C. and J. W. Ekin. “Large intrinsic effect of axial strain on the critical current of high-temperature superconductors for electric power applications”. In: *APPLIED PHYSICS LETTERS* 90.5 (2007), p. 52506. DOI: 10.1063/1.2435612.

[vGH12] van der Laan, D. C., L. F. GOODRICH and T. J. Haugan. “High-current dc power transmission in flexible RE–Ba2Cu3O7 $-\ \delta$ coated conductor cables”. In: *Superconductor Science and Technology* 25.1 (2012), p. 14003. DOI: 10.1088/0953-2048/25/1/014003.

[van+10] van der Laan, D. C. et al. “Effect of strain, magnetic field and field angle on the critical current density of Y Ba2Cu3O7$-\delta$ coated conductors”. In: *Superconductor Science and Technology* 23.7 (2010), p. 72001. DOI: 10.1088/0953-2048/23/7/072001.

[van+13] van der Laan, D. C. et al. “Characterization of a high-temperature superconducting conductor on round core cables in magnetic fields up to 20 T”. in: *Superconductor Science and Technology* 26.4 (2013), p. 45005. DOI: 10.1088/0953-2048/26/4/045005.

[VM95] G. VELLEGO and P. METRA. “AN ANALYSIS OF THE TRANSPORT LOSSES MEASURED ON HTSC SINGLE-PHASE CONDUCTOR PROTOTYPES”. in: *SUPERCONDUCTOR SCIENCE & TECHNOLOGY* 8.6 (1995), pp. 476–483.

[Vla+18] V. R. Vlad et al. “Inkjet Printing Multideposited YBCO on CGO/LMO/MgO/Y2O3/Al2 O3/Hastelloy Tape for 2G-Coated Conductors”. In:

IEEE TRANSACTIONS ON APPLIED SUPERCONDUCTIVITY 28.4 (2018), pp. 1–5. DOI: 10.1109/TASC.2018.2808403.

[Vys+13] V. S. Vysotsky et al. "Hybrid Energy Transfer Line With Liquid Hydrogen and Superconducting MgB2 Cable-First Experimental Proof of Concept". In: *IEEE TRANSACTIONS ON APPLIED SUPERCONDUCTIVITY* 23.3 (2013). DOI: 10.1109/TASC.2013.2238574.

[Vys+15] V. S. Vysotsky et al. "New 30-m flexible hybrid energy transfer line with liquid hydrogen and superconducting MgB2 cable - Development and test results". In: *IEEE TRANSACTIONS ON APPLIED SUPERCONDUCTIVITY* 25.3 (2015). DOI: 10.1109/TASC.2014.2361635.

[Wat+17] H. Watanabe et al. "Cooling test of the 500 m class superconducting DC power transmission system". In: *IOP Conference Series: Materials Science and Engineering* 171.1 (2017), p. 012116. DOI: 10.1088/1757-899X/171/1/012116.

[Wei17] J. Weiss. *Current progress of High-temperature superconducting CORC® magnet cable and wire development.* Geneva, 2017. URL: https://indico.cern.ch/event/659554/contributions/2709426/attachments/1527211/2388338/3LO2-04_Jeremy_Weiss_Room_1.pdf.

[WC84] G. K. WHITE and S. J. COLLOCOTT. "HEAT-CAPACITY OF REFERENCE MATERIALS - CU AND W". in: *Journal of Physical and Chemical Reference Data* 13.4 (1984), pp. 1251–1257.

[WFP17] M. J. Wolf, W. H. Fietz and A. Preuss. "Investigation of HTS CrossConductor Joints, Connectors, and Terminations". In: *IEEE TRANSACTIONS ON APPLIED SUPERCONDUCTIVITY* 27.4 (2017), p. 4802605. DOI: 10.1109/TASC.2017.2652305.

[Wol+16] M. J. Wolf et al. "HTS CroCo: A Stacked HTS Conductor Optimized for High Currents and Long-Length Production". In: *IEEE TRANSACTIONS ON APPLIED SUPERCONDUCTIVITY* 26.2 (2016), p. 6400106. DOI: 10.1109/TASC.2016.2521323.

[WU+87] M. K. WU et al. "SUPERCONDUCTIVITY AT 93-K IN A NEW MIXED-PHASE Y-BA-CU-O COMPOUND SYSTEM AT AMBIENT PRESSURE". in: *PHYSICAL REVIEW LETTERS* 58.9 (1987), pp. 908–910.

[Yam+15] S. Yamaguchi et al. "Concept and design of 500 meter and 1000 meter dc superconducting power cables in Ishikari, Japan". In: *IEEE TRANSACTIONS ON APPLIED SUPERCONDUCTIVITY* 25.3 (2015), p. 5402504. DOI: 10.1109/TASC.2015.2390045.

[Yan+15] B. Yang et al. "Qualification Test of a 80 kV 500 MW HTS DC Cable for Applying Into Real Grid". In: *IEEE TRANSACTIONS ON APPLIED SUPERCONDUCTIVITY* 25.3 (2015), p. 5402705. DOI: 10.1109/TASC.2015.2396683.

[Yaz+13] S. Yazaki et al. "Critical Current Degradation in High-Temperature Superconducting Tapes Caused by Temperature Rise". In: *IEEE TRANSACTIONS ON APPLIED SUPERCONDUCTIVITY* 23.3 (2013), p. 4602304. DOI: 10.1109/TASC.2013.2244157.

[Yoo+16] S. Yoon et al. "26 T 35 mm all-GdBa2Cu3O7−x multi-width no-insulation superconducting magnet". In: *Superconductor Science and Technology* 29.4 (2016), 04LT04. DOI: 10.1088/0953-2048/29/4/04LT04. URL: https://iopscience.iop.org/article/10.1088/0953-2048/29/4/04LT04/pdf.

[Yos+14] K. Yoshitomi et al. "AC Loss of Ripple Current in Superconducting DC Power Transmission Cable". In: *Physics Procedia* 58 (2014), pp. 326–329. DOI: 10.1016/j.phpro.2014.09.080.

[Yum+09] H. Yumura et al. "Phase II of the Albany HTS Cable Project". In: *IEEE TRANSACTIONS ON APPLIED SUPERCONDUCTIVITY* 19.3 (2009), pp. 1698–1701. DOI: 10.1109/TASC.2009.2017865.

[ZQG16] V. M. R. Zermeño, S. Quaiyum and F. Grilli. "Open-Source Codes for Computing the Critical Current of Superconducting Devices". In: *IEEE TRANSACTIONS ON APPLIED SUPERCONDUCTIVITY* 26.3 (2016), pp. 1–7. DOI: 10.1109/TASC.2016.2521171.

[Zer12] U. Zerweck-Trogisch. *Förderpumpen für kryogene Flüssigkeiten.* 5.10.2012. URL: https://www.ilkdresden.de/projekt/kryofluessigkeitspumpe-fuer-ln2-h2-lng/.

[Zha+15] D. Zhang et al. "Stability Analysis of the Cable Core of a 10 kA HTS DC Power Cable Used in the Electrolytic Aluminum Industry". In: *IEEE TRANSACTIONS ON APPLIED SUPERCONDUCTIVITY* 25.3 (2015). DOI: 10.1109/TASC.2014.2374691.

[Zha+11] Y. Zhang et al. "Solderability study of RABiTS-based 5YBCO6 coated conductors". In: *Physica C: Superconductivity* 471.15–16 (2011), pp. 437–443. DOI: 10.1016/j.physc.2011.03.009.

Karlsruher Schriftenreihe zur Supraleitung

Karlsruher Institut für Technologie (KIT) | ISSN 1869-1765

Band 008 **Axel Stockhausen**
Optimization of Hot-Electron Bolometers for THz Radiation. 2013
ISBN 978-3-7315-0066-7

Band 009 **Petra Thoma**
Ultra-fast $YBa_2Cu_3O_{7-x}$ direct detectors for the THz frequency range. 2013
ISBN 978-3-7315-0070-4

Band 010 **Dagmar Henrich**
Influence of Material and Geometry on the Performance of Superconducting Nanowire Single-Photon Detectors. 2013
ISBN 978-3-7315-0092-6

Band 011 **Alexander Scheuring**
Ultrabreitbandige Strahlungseinkopplung in THz-Detektoren. 2013
ISBN 978-3-7315-0102-2

Band 012 **Markus Rösch**
Development of lumped element kinetic inductance detectors for mm-wave astronomy at the IRAM 30 m telescope. 2013
ISBN 978-3-7315-0110-7

Band 013 **Johannes Maximilian Meckbach**
Superconducting Multilayer Technology for Josephson Devices. 2013
ISBN 978-3-7315-0122-0

Band 014 **Enrico Rizzo**
Simulations for the optimization of High Temperature Superconductor current leads for nuclear fusion applications. 2014
ISBN 978-3-7315-0132-9

Band 015 **Philipp Krüger**
Optimisation of hysteretic losses in high-temperature superconducting wires. 2014
ISBN 978-3-7315-0185-5

Band 016 **Matthias Hofherr**
Real-time imaging systems for superconducting nanowire single-photon detector arrays. 2014
ISBN 978-3-7315-0229-6

Band 017 **Oliver Näckel**
Development of an Air Coil Superconducting Fault Current Limiter. 2016
ISBN 978-3-7315-0526-6

Band 018 **Christoph M. Bayer**
Characterization of High Temperature Superconductor Cables for Magnet Toroidal Field Coils of the DEMO Fusion Power Plant. 2017
ISBN 978-3-7315-0605-8

Band 019 **Shengnan Zou**
Magnetization of High Temperature Superconducting Trapped-Field Magnets. 2017
ISBN 978-3-7315-0715-4

Band 020 **Ilya Charaev**
Improving the Spectral Bandwidth of Superconducting Nanowire Single-Photon Detectors (SNSPDs). 2018
ISBN 978-3-7315-0745-1

Band 021 **Juliane Raasch**
Electrical-field sensitive $YBa_2Cu_3O_{7-x}$ detectors for real-time monitoring of picosecond THz pulses. 2019
ISBN 978-3-7315-0786-4

Band 022 **Yingzhen Liu**
Design of a superconducting DC wind generator. 2020
ISBN 978-3-7315-0796-3

Band 023 **Sebastian Hellmann**
Research and Technology Development on Superconducting Current Limiting Transformers. 2019
ISBN 978-3-7315-0804-5

Band 024 **Simon J. Otten**
Characterisation of REBCO Roebel cables. 2019
ISBN 978-3-7315-0904-2

Band 025 **Julia Brandel**
Supraleitende Einzelphotonenzähler: Optimierung der Zeitauflösung und Anwendungsbeispiele aus der Spektroskopie. 2019
ISBN 978-3-7315-0917-2

Band 026 **Dustin Kottonau, Eugen Shabagin, Wescley T. B. de Sousa, Jörn Geisbüsch, Mathias Noe, Hanno Stagge, Simon Fechner, Hannes Woiton, Thomas Küsters**
Bewertung des Einsatzes supraleitender 380-kV-Kabel. 2019
ISBN 978-3-7315-0927-1

Band 027 **Steffen Dörner**
Multifrequenzausleseverfahren von supraleitenden Einzelphotonen-Detektoren. 2019
ISBN 978-3-7315-0961-5

Band 028 **Michael Merker**
Superconducting integrated THz receiver. 2019
ISBN 978-3-7315-0970-7

Band 029 **Wolfgang-Gustav Ekkehart Schmidt**
Superconducting Nanowire Single-Photon Detectors for Quantum Photonic Integrated Circuits on GaAs. 2020
ISBN 978-3-7315-0980-6

Band 030 **Dustin Kottonau, Eugen Shabagin, Wescley de Sousa, Jörn Geisbüsch, Mathias Noe, Hanno Stagge, Simon Fechner, Hannes Woiton, Thomas Küsters**
Evaluation of the Use of Superconducting 380 kV Cable. 2020
ISBN 978-3-7315-1026-0

Band 031 **Alan Preuß**
Development of high-temperature superconductor cables for high direct current applications. 2021
ISBN 978-3-7315-1041-3

Die Bände sind unter www.ksp.kit.edu als PDF frei verfügbar oder als Druckausgabe bestellbar.